T0291815

# CAMBRIDGE LIBRARY COLLECTION

*Books of enduring scholarly value*

## Mathematics

From its pre-historic roots in simple counting to the algorithms powering modern desktop computers, from the genius of Archimedes to the genius of Einstein, advances in mathematical understanding and numerical techniques have been directly responsible for creating the modern world as we know it. This series will provide a library of the most influential publications and writers on mathematics in its broadest sense. As such, it will show not only the deep roots from which modern science and technology have grown, but also the astonishing breadth of application of mathematical techniques in the humanities and social sciences, and in everyday life.

## The Elements of Algebra

In his autobiography, Charles Darwin wrote of his time at Cambridge: 'I attempted mathematics … but I got on very slowly. The work was repugnant to me, chiefly from my not being able to see any meaning in the early steps in algebra. This impatience was very foolish, and in after years I have deeply regretted that I did not proceed far enough at least to understand something of the great leading principles of mathematics, for men thus endowed seem to have an extra sense.' First published in 1795 and reissued here in its 1815 sixth edition, *The Elements of Algebra* by James Wood (1760–1839) was one of the standard Cambridge texts for decades, so its presence in Darwin's library aboard the *Beagle* is readily understandable. Then, as now, Cambridge had a high opinion of itself as a mathematical university. The contents of Wood's book give an interesting glimpse of the standards expected of the less able students.

Cambridge University Press has long been a pioneer in the reissuing of out-of-print titles from its own backlist, producing digital reprints of books that are still sought after by scholars and students but could not be reprinted economically using traditional technology. The Cambridge Library Collection extends this activity to a wider range of books which are still of importance to researchers and professionals, either for the source material they contain, or as landmarks in the history of their academic discipline.

Drawing from the world-renowned collections in the Cambridge University Library and other partner libraries, and guided by the advice of experts in each subject area, Cambridge University Press is using state-of-the-art scanning machines in its own Printing House to capture the content of each book selected for inclusion. The files are processed to give a consistently clear, crisp image, and the books finished to the high quality standard for which the Press is recognised around the world. The latest print-on-demand technology ensures that the books will remain available indefinitely, and that orders for single or multiple copies can quickly be supplied.

The Cambridge Library Collection brings back to life books of enduring scholarly value (including out-of-copyright works originally issued by other publishers) across a wide range of disciplines in the humanities and social sciences and in science and technology.

# The Elements of Algebra

*Designed for the Use of Students in the University*

JAMES WOOD

CAMBRIDGE
UNIVERSITY PRESS

# CAMBRIDGE
## UNIVERSITY PRESS

University Printing House, Cambridge, CB2 8BS, United Kingdom

Published in the United States of America by Cambridge University Press, New York

Cambridge University Press is part of the University of Cambridge.

It furthers the University's mission by disseminating knowledge in the pursuit of education, learning and research at the highest international levels of excellence.

www.cambridge.org
Information on this title: www.cambridge.org/9781108066532

© in this compilation Cambridge University Press 2014

This edition first published 1815
This digitally printed version 2014

ISBN 978-1-108-06653-2 Paperback

# Titles known to have formed part of Charles Darwin's library during the *Beagle* voyage, available in the
# CAMBRIDGE LIBRARY COLLECTION

Abel, Clarke: *Narrative of a Journey in the Interior of China, and of a Voyage to and from that Country in the Years 1816 and 1817* (1818) [ISBN 9781108045995]

Aubuisson de Voisins, J.F. d': *Traité de Géognosie* (2 vols., 1819) [ISBN 9781108029728]

Bougainville, L. de, translated by John Reinhold Forster: *A Voyage Round the World, Performed by Order of His Most Christian Majesty, in the Years 1766–1769* (1772) [9781108031875]

Buch, Leopold von, translated by John Black, with notes and illustrations by Robert Jameson: *Travels through Norway and Lapland during the years 1806, 1807, and 1808* (1813) [ISBN 9781108028813]

Byron, John: *The Narrative of the Honourable John Byron, Commodore in a Late Expedition Round the World* (1768) [ISBN 9781108065368]

Caldcleugh, Alexander: *Travels in South America, during the Years, 1819–20–21* (2 vols., 1825) [ISBN 9781108033732]

Callcott, Maria (née Graham): *Voyage of H.M.S. Blonde to the Sandwich Islands, in the Years 1824–1825* (1826) [ISBN 9781108062114]

Candolle, Augustin Pyramus de, and Sprengel, Kurt: *Elements of the Philosophy of Plants* (1821) [ISBN 9781108037464]

Colnett, James: *A Voyage to the South Atlantic and Round Cape Horn into the Pacific Ocean* (1798) [ISBN 9781108048354]

Cuvier, Georges: *Le règne animal distribué d'après son organisation* (4 vols., 1817) [ISBN 9781108058872]

Cuvier, Georges, edited by Edward Griffith: *The Animal Kingdom* (16 vols., 1827–35) [ISBN 9781108049702]

Daniell, J. Frederic: *Meteorological Essays and Observations* (1827) [ISBN 9781108056571]

De la Beche, Henry T.: *A Selection of the Geological Memoirs Contained in the Annales des Mines* (1824) [ISBN 9781108048408]

Earle, Augustus: *A Narrative of a Nine Months' Residence in New Zealand in 1827* (1832) [ISBN 9781108039789]

Ellis, William: *Polynesian Researches during a Residence of Nearly Six Years in the South Sea Islands* (2 vols., 1829) [ISBN 9781108065382]

Falkner, Thomas: *A Description of Patagonia, and the Adjoining Parts of South America* (1774) [ISBN 9781108060547]

Fleming, John: *The Philosophy of Zoology* (2 vols., 1822) [ISBN 9781108001649]

Flinders, Matthew: *A Voyage to Terra Australis* (2 vols., 1814) [ISBN 9781108018203]

Forster, John Reinhold: *Observations Made During a Voyage Round the World* (1778) [ISBN 9781108031882]

Greenough, George Bellas: *Critical Examination of the First Principles of Geology* (1819) [ISBN 9781108035323]

Hawkesworth, John: *An Account of the Voyages Undertaken by the Order of His Present Majesty for Making Discoveries in the Southern Hemisphere* (3 vols., 1773) [ISBN 9781108065528]

Head, Francis Bond: *Rough Notes Taken during some Rapid Journeys across the Pampas and among the Andes* (1826) [ISBN 9781108001618]

Humboldt, Alexander von: *Essai géognostique sur le gisement des roches dans les deux hémisphères* (1826) [ISBN 9781108049481]

Humboldt, Alexander von, translated by J.B.B. Eyriès: *Tableaux de la nature* (1828) [ISBN 9781108052757]

Humboldt, Alexander von, translated by Helen Maria Williams: *Personal Narrative of Travels* (7 vols., 1814–29) [ISBN 9781108028004]

Humboldt, Alexander von: *Fragmens de géologie et de climatologie Asiatiques* (2 vols., 1831) [ISBN 9781108049443]

Jones, Thomas: *A Companion to the Mountain Barometer* (1817) [ISBN 9781108049375]

King, Phillip Parker: *Narrative of a Survey of the Intertropical and Western Coasts of Australia, Performed between the Years 1818 and 1822* (2 vols., 1827) [ISBN 9781108045988]

Kirby, William and Spence, William: *An Introduction to Entomology* (4 vols., 1815–26) [ISBN 9781108065597]

Kotzebue, Otto von, translated by H.E. Lloyd: *A Voyage of Discovery, into the South Sea and Beering's Straits, for the Purpose of Exploring a North-East Passage* (3 vols., 1821) [ISBN 9781108057608]

Lamarck, Jean-Baptiste Pierre Antoine de Monet de: *Histoire naturelle des animaux sans vertèbres* (7 vols., 1815–22) [ISBN 9781108059084]

Lyell, Charles: *Principles of Geology* (3 vols., 1830–3) [ISBN 9781108001342]

Macdouall, John: *Narrative of a Voyage to Patagonia and Terra del Fuego* (1833) [ISBN 9781108060981]

Mawe, John: *Travels in the Interior of Brazil* (1821) [ISBN 9781108052788]

Miers, John: *Travels in Chile and La Plata* (2 vols., 1826) [ISBN 9781108072977]

Molina, Giovanni Ignazio: *The Geographical, Natural, and Civil History of Chili* (2 vols., 1782–6, English translation 1809) [ISBN 9781108049474]

Owen, William Fitzwilliam, translated by Heaton Bowstead Robinson: *Narrative of Voyages to Explore the Shores of Africa, Arabia, and Madagascar* (2 vols., 1833) [ISBN 9781108050654]

Pernety, Antoine-Joseph: *The History of a Voyage to the Malouine (or Falkland) Islands* (1770, English translation 1771) [ISBN 9781108064330]

Phillips, William: *An Elementary Introduction to the Knowledge of Mineralogy* (1816) [ISBN 9781108049382]

Playfair, John: *Illustrations of the Huttonian Theory of the Earth* (1802) [ISBN 9781108072311]

Scrope, George Poulett: *Considerations on Volcanos* (1825) [ISBN 9781108072304]

Southey, Robert: *History of Brazil* (3 vols., 1810–19) [ISBN 9781108052870]

Spix, Johann Baptist von, and Martius, C.F.P. von, translated by H.E. Lloyd: *Travels in Brazil, in the Years 1817–1820* (2 vols., 1824) [ISBN 9781108063807]

Turnbull, John: *A Voyage Round the World, in the Years 1800, 1801, 1802, 1803, and 1804* (1805, this edition 1813) [ISBN 9781108053983]

Ulloa, Antonio de, translated and edited by John Adams: *A Voyage to South America* (2 vols., 1806) [ISBN 9781108031707]

Volney, Constantin-François: *Voyage en Syrie et en Égypte pendant les années 1783, 1784 et 1785* (2 vols., 1787) [ISBN 9781108066556]

Webster, William Henry Bayley: *Narrative of a Voyage to the Southern Atlantic Ocean, in the Years 1828, 29, 30, Performed in H.M. Sloop Chanticleer* (2 vols., 1834) [ISBN 9781108041898]

Weddell, James: *A Voyage towards the South Pole: Performed in the Years 1822–24* (1825) [ISBN 9781108041584]

Wood, James: *The Elements of Algebra* (1815) [ISBN 9781108066532]

For a complete list of titles in the Cambridge Library Collection please visit: www.cambridge.org/features/CambridgeLibraryCollection/books.htm

THE

# ELEMENTS OF ALGEBRA:

DESIGNED

## FOR THE USE OF STUDENTS

IN THE

## UNIVERSITY.

———◆———

## BY JAMES WOOD, B.D.

FELLOW OF ST. JOHN'S COLLEGE, CAMBRIDGE.

**SIXTH EDITION.**

CAMBRIDGE:

*Printed by J. Smith, Printer to the University;*

AND SOLD BY J. DEIGHTON, AND J. NICHOLSON, CAMBRIDGE;
AND W. H. LUNN, SOHO-SQUARE, LONDON.

———

1815.

# CONTENTS.

# CONTENTS.

INTRODUCTION.

# INTRODUCTION.

## ON VULGAR FRACTIONS.

ARTICLE 1. A *fraction* is a quantity which represents a part or parts of an integer or whole.

(2.) A *simple fraction* consists of two members, the *numerator* and the *denominator*; the denominator shews into how many equal parts the whole, or unity, is divided; and the numerator, the number of those parts taken. The numerator is usually placed over the denominator with a line between them. Thus, $\frac{2}{3}$, two thirds, signifies that unity is divided into three equal parts, and that two of those parts are taken.

It must be observed, that we suppose every integer to be divisible into any number of equal parts at pleasure.

(3.) A *proper fraction* is one whose numerator is less than it's denominator, as $\frac{7}{8}$.

(4.) An *improper fraction* is one whose numerator is equal to, or greater than it's denominator, as $\frac{6}{6}$; $\frac{7}{5}$

A                                             (5.) A

(5.) A *compound fraction* is a fraction of a fraction, as $\frac{3}{4}$ of $\frac{5}{6}$, where $\frac{5}{6}$ is the whole quantity of which $\frac{3}{4}$ is to be taken; also, $\frac{2}{3}$ of $\frac{4}{5}$ of $\frac{9}{11}$, is a compound fraction; &c.

(6.) A quantity consisting of a whole number and a fraction is called a *mixed number*, as $7\frac{3}{10}$, which signifies 7 integers together with $\frac{3}{10}$ of an integer.

(7.) Cor. 1. Every integer may be considered as a fraction whose denominator is 1; thus 5, or five units, is $\frac{5}{1}$.

(8.) Cor. 2. *To multiply a fraction by any number, multiply the numerator by that number and retain the same denominator.* Thus, $\frac{2}{15}$ multiplied by 7 is $\frac{14}{15}$. For, the unit, in each of the fractions $\frac{2}{15}$ and $\frac{14}{15}$, is divided into 15 equal parts, and 7 times as many of those parts are taken in the latter case as in the former.

(9.) Cor. 3. *To divide a fraction by any number, multiply the denominator by that number and retain the same numerator.* Thus, $\frac{3}{5}$ divided by 4 is $\frac{3}{20}$. For, the unit being divided into four times as many equal parts in $\frac{3}{20}$ as it is in $\frac{3}{5}$, each of the parts in the latter case is four times as great as in the former, and the

the same number of parts is taken in both cases; therefore the former fraction is one fourth of the latter.

(10.) A simple fraction may be considered as representing the quotient arising from the division of the numerator by the denominator; thus the fraction $\frac{3}{4}$ represents the quotient of 3 divided by 4; for 3 is $\frac{3}{1}$ (Art. 7), and this divided by 4 is the fraction $\frac{3}{4}$ (Art. 9). If the integer be supposed a pound, or twenty shillings, $\frac{3}{4}$ of £.1, which is 15 shillings, is equal to $\frac{1}{4}$ of £.3, which is also 15 shillings.

(11.) If the numerator and denominator of a fraction be both multiplied by the same number, it's value is not altered. For, if the numerator be multiplied by any number, the fraction is multiplied by that number (Art. 8); and if the denominator be multiplied by the same number, the fraction is divided by it (Art. 9); and if a quantity be both multiplied and divided by the same number, it's value is not altered. Thus, $\frac{5}{14} = \frac{15}{42} = \frac{150}{420}$, &c. Hence, if the numerator and denominator be both divided by the same number, it's value is not altered; since $\frac{150}{420} = \frac{15}{42} = \frac{5}{14}$ *.

---

* To avoid repetition, the Reader is referred to the first section of the Algebra, for the explanation of the signs ˥ , ─, ×, and ═.

ON

## ON REDUCTION.

The operation by which a quantity is changed from one denomination to another, without altering it's value, is called *Reduction*.

(12.) *To reduce a whole number to a fraction with a given denominator.*

Multiply the proposed number by the given denominator, and the product will be the numerator of the fraction required.

Ex. Reduce 5 to a fraction whose denominator is 6.

This is $\dfrac{5 \times 6}{6}$ or $\dfrac{30}{6}$; because 5 may be considered

as a fraction $\dfrac{5}{1}$ (Art. 7), the numerator and denominator of which are multiplied by 6, therefore it's value is not altered. (Art. 11).

(13.) *To reduce a mixed number to an improper fraction.*

Multiply the integral by the denominator of the fractional part, to this product add the numerator of the fractional part, and make it's denominator the denominator of the sum.

Ex. 1. Reduce $7\dfrac{4}{5}$ to an improper fraction.

The quantity $7\dfrac{4}{5}$ is equal to $\dfrac{35+4}{5} = \dfrac{39}{5}$; for 7

(by the last Art.) is equal to $\dfrac{35}{5}$, and if to this $\dfrac{4}{5}$ be

added, the whole is $\dfrac{39}{5}$.

Ex. 2. Also, $23\dfrac{9}{11} = \dfrac{253+9}{11} = \dfrac{262}{11}$.

(14.) *To*

(14.) *To reduce an improper fraction to a mixed number.*

Divide the numerator by the denominator for the integral part, and make the remainder the numerator of the fractional part, and the divisor it's denominator.

Ex. Reduce $\frac{39}{5}$ to an improper fraction. The fraction $\frac{39}{5} = 7\frac{4}{5}$ ; because the unit being divided into 5 parts, 39 such parts are to be taken, that is, 7 units and 4 such parts.

(15.) *To reduce a compound fraction to a simple one.*

Multiply all the numerators together for a new numerator, and all the denominators for a new denominator.

Ex. 1. $\frac{2}{3}$ of $\frac{4}{5} = \frac{8}{15}$ ; for, one third of $\frac{4}{5}$ is $\frac{4}{15}$ , (Art. 9); therefore two thirds, which must be twice as great, is $\frac{8}{15}$ (Art. 8).

Ex. 2. $\frac{3}{4}$ of $5 = \frac{3}{4}$ of $\frac{5}{1} = \frac{15}{4}$ .

Mixed numbers must be reduced to improper fractions, before the rule can be applied.

Ex. 3. $\frac{5}{8}$ of $\frac{2}{9}$ of $3\frac{1}{12} = \frac{10}{72}$ of $3\frac{1}{12} = \frac{10}{72}$ of $\frac{37}{12}$

$= \frac{370}{864}$ .

(16.) *To reduce a fraction to lower terms.*

Whenever the numerator and denominator of a fraction have a common measure (or number which divides each of them without remainder) greater than unity.

unity, the fraction may be reduced to lower terms, by dividing both the numerator and denominator by this common measure.

Ex. $\frac{105}{120}$ is reduced to $\frac{21}{24}$, by dividing both the numerator and denominator by 5; and $\frac{21}{24}$ is again reduced to $\frac{7}{8}$, by dividing it's numerator and denominator by 3. That the value of the fraction is not altered, appears from Art. 11.

In the same manner, $\frac{168}{210} = \frac{84}{105} = \frac{28}{35} = \frac{4}{5}$.

(17.) *The greatest common measure of two numbers is found by dividing the greater by the less, and the preceding divisor by the remainder, continually, till nothing is left. the last divisor is the greatest common measure required.*

To find the greatest common measure of 189 and 224.

$$189)224(1$$
$$189$$
$$\overline{\phantom{00}}$$
$$35)189(5$$
$$175$$
$$\overline{\phantom{00}}$$
$$14)35(2$$
$$28$$
$$\overline{\phantom{00}}$$
$$7)14(2$$
$$14$$
$$\overline{\phantom{0}*}$$

By proceeding according to the rule, it appears that 7 is

is the last divisor, or the greatest common measure sought. The proof of this rule will be given hereafter*.

(18.) *A fraction is reduced to it's* lowest *terms, by dividing it's numerator and denominator by their* greatest *common measure.*

Ex. To reduce $\dfrac{385}{396}$ to it's lowest terms.

By the last Art. the greatest common measure of the numerator and denominator is found to be 11, and therefore $\dfrac{35}{36}$ is the fraction in it's lowest terms.

Cor. If unity be the greatest common measure of the numerator and denominator, the fraction is in it's lowest terms.

(19.) *To reduce fractions to a common denominator.*

Having reduced, if necessary, compound fractions to simple ones, and mixed numbers to improper fractions, multiply each numerator by all the denominators except it's own, for the new numerator, and all the denominators together for a common denominator.

Ex. 1. Reduce $\dfrac{1}{2}$, $\dfrac{2}{3}$ and $\dfrac{3}{4}$ to a common denominator.

$$\dfrac{1\times3\times4}{2\times3\times4}, \dfrac{2\times2\times4}{2\times3\times4}, \text{ and } \dfrac{3\times2\times3}{2\times3\times4}, \text{ or } \dfrac{12}{24}, \dfrac{16}{24} \text{ and } \dfrac{18}{24}$$

are the fractions required. These fractions are respectively equal to the former, the numerator and denominator in each case, having been multiplied by the same numbers, namely, the denominators of the rest.

$$\dfrac{1\times3\times4}{2\times3\times4}=\dfrac{1}{2}; \dfrac{2\times2\times4}{2\times3\times4}=\dfrac{2}{3}; \text{ and } \dfrac{3\times2\times3}{2\times3\times4}=\dfrac{3}{4}.$$

Ex. 2.

* Art. 90.

Ex. 2. Reduce $\frac{2}{5}$ of $\frac{3}{4}$ and $4\frac{1}{3}$ to a common denominator.

These are $\frac{6}{20}$ and $\frac{13}{3}$, or $\frac{3}{10}$ and $\frac{13}{3}$; therefore $\frac{9}{30}$ and $\frac{130}{30}$ are the fractions required.

(20.) *If the denominator of one of two fractions contain the denominator of the other a certain number of times exactly, multiply the numerator and denominator of the latter by that number, and it will be reduced to the same denominator with the former.*

Ex. Reduce $\frac{5}{12}$ and $\frac{2}{3}$ to a common denominator.

Since 12 contains 3 four times exactly, multiply both the numerator and denominator of $\frac{2}{3}$ by 4, and it becomes $\frac{8}{12}$, a fraction having the same denominator with $\frac{5}{12}$.

(21.) Cor. By reducing two fractions to a common denominator their values may be compared.

Thus, $\frac{4}{7}$ and $\frac{7}{12}$ when reduced to a common denominator are $\frac{48}{84}$ and $\frac{49}{84}$; that is, the fractions have the same relative values that 48 and 49 have.

(22.) *To find the value of a fraction of a proposed denomination in terms of a lower denomination.*

Multiply

Multiply the fraction by the number of integers of the lower denomination contained in one integer of the higher, and the product is the value required. The value of any fractional part of the lower denomination may be obtained in the same manner, till we come to the lowest.

Ex. 1. What is the value of $\dfrac{5}{7}$ of a pound?

First, $\dfrac{5}{7}$ of £. 1 is $\dfrac{5}{7}$ of 20 shillings, or $\dfrac{5}{7}$ of $\dfrac{20}{1}$ shillings $= \dfrac{100}{7} = 14\dfrac{2}{7}$ shillings;

Next, $\dfrac{2}{7}$ of a shilling $= \dfrac{2}{7}$ of $\dfrac{12}{1}$ pence $= \dfrac{24}{7}$ pence $= 3\dfrac{3}{7}$ pence;

Lastly, $\dfrac{3}{7}$ of a penny $= \dfrac{3}{7}$ of 4 farthings $= \dfrac{3}{7}$ of $\dfrac{4}{1}$, or $\dfrac{12}{7}$ farthings $= 1\dfrac{5}{7}$ farthings: hence, $\dfrac{5}{7}$ of a pound is $\overset{s.}{14} : \overset{d.}{3} : \overset{q.}{1}\dfrac{5}{7}$.

The operation is usually performed in the following manner:

$$\pounds.5$$
$$20$$
$$\overline{\phantom{0}}$$
$$7)100$$
$$\overline{\phantom{0}}$$
$$14 - - 2s.$$
$$12$$
$$\overline{\phantom{0}}$$
$$7)24$$
$$\overline{\phantom{0}}$$
$$3 - - 3d.$$
$$4$$
$$\overline{\phantom{0}}$$
$$7)12$$
$$\overline{\phantom{0}}$$
$$1 - - 5q.$$

Ans. $\overset{s.}{14} : \overset{d.}{3} : \overset{q.}{1}\dfrac{5}{7}$.

Ex. 2. What is the value of $\dfrac{5}{9}$ of a crown?

$$5 \; C$$
$$5$$
$$\overline{\phantom{0}}$$
$$9)25$$
$$\overline{\phantom{0}}$$
$$2 - - 7s.$$
$$12$$
$$\overline{\phantom{0}}$$
$$9)84$$
$$\overline{\phantom{0}}$$
$$9 - - 3d.$$
$$4$$
$$\overline{\phantom{0}}$$
$$9)12$$
$$\overline{\phantom{0}}$$
$$1 - - 3q.$$

Ans. $\overset{s.}{2} : \overset{d.}{9} : \overset{q.}{1}\dfrac{3}{9}$.

(23.) *To*

(23.) *To reduce a quantity to a fraction of any denomination.*

Make the given quantity the numerator, and the number of integers of it's denomination in one of the proposed denomination, the denominator, and the fraction required is determined.

Ex. 1. What fraction of a pound is $\overset{s.}{12} : \overset{d.}{7} : \overset{q.}{3}$ ?

$\overset{s.}{12} : \overset{d.}{7} : \overset{q.}{3} = \overset{q.}{607}$, and one pound $= \overset{q.}{960}$; therefore $\frac{607}{960}$ is the fraction sought. Because the integer being divided into 960 equal parts, $\overset{s.}{12} : \overset{d.}{7} : \overset{q.}{3}$ contains 607 such parts.

(24.) In this example, we are obliged to reduce the whole to farthings; and in general, if the higher denomination do not contain the lower an exact number of times, reduce them to a common denomination, and proceed as before.

Ex. 2. What fraction of a guinea is half a crown?

Here sixpence is the greatest common denomination, of which a guinea contains 42, and half a crown 5, therefore $\frac{5}{42}$ is the fraction required.

Any common denomination would answer the purpose, but if the greatest be taken, the resulting fraction is in the lowest terms.

(25.) *To reduce a fraction to any denomination.*

Find what fraction of the proposed denomination an integer of the denomination of the given fraction is, and the fraction required will be found by Art. 15.

Ex. 1.

Ex. 1. What fraction of a pound is $\frac{2}{3}$ of a shilling?

1 shilling is $\frac{1}{20}$ of a pound, therefore $\frac{2}{3}$ of 1 shilling

is $\frac{2}{3}$ of $\frac{1}{20}$ of a pound, or $\frac{2}{60} = \frac{1}{30}$ of a pound.

Ex. 2. What fraction of a yard is $\frac{5}{7}$ of an inch?

1 inch is $\frac{1}{36}$ of a yard, therefore $\frac{5}{7}$ of an inch is $\frac{5}{7}$

of $\frac{1}{36}$ of a yard, or $\frac{5}{252}$ of a yard.

Ex. 3. What fraction of a guinea is $\frac{4}{9}$ of a pound?

1 pound is $\frac{20}{21}$ of a guinea (Art. 24); hence, $\frac{4}{9}$ of a

pound is $\frac{4}{9}$ of $\frac{20}{21}$ of a guinea, or $\frac{80}{189}$ of a guinea.

## ADDITION OF FRACTIONS.

(26.) *If fractions have a common denominator, their sum is found by taking the sum of the numerators, and subjoining the common denominator.*

Ex. $\frac{1}{5} + \frac{2}{5} = \frac{3}{5}$. For, if an integer be divided into five equal parts, one of those parts, together with two parts of the same kind, must make three such parts.

(27.) If the fractions have not a common denominator, reduce them to a common denominator, and proceed as before.

Ex.

Ex. Required the sum of $\frac{2}{3}$, $\frac{3}{4}$ and $\frac{4}{5}$.

These reduced to a common denominator are $\frac{40}{60}$, $\frac{45}{60}$ and $\frac{48}{60}$, whose sum is $\frac{133}{60}$, or $2\frac{13}{60}$.

When mixed numbers are to be added, to the sum of the fractions, taken as before, add the sum of the integers.

Ex. Add together $5\frac{3}{4}$, $6\frac{1}{3}$ and $\frac{2}{5}$ of $\frac{1}{7}$.

$$\frac{3}{4} + \frac{1}{3} + \frac{2}{35} = \frac{315}{420} + \frac{140}{420} + \frac{24}{420} = \frac{479}{420} = 1\frac{59}{420};$$

therefore the whole sum is $12\frac{59}{420}$.

## SUBTRACTION.

(28.) *The difference of two fractions which have a common denominator is found by taking the difference of their numerators and subjoining the common denominator.*

Ex. $\frac{4}{5} - \frac{3}{5} = \frac{1}{5}$. For, if the unit be supposed to be divided into five equal parts, and three of those parts be taken from four, the remainder must be one, or $\frac{1}{5}$.

(29.) If the fractions have not a common denominator, let them be reduced to a common denominator, and take the difference as before.

Ex.

Ex. 1.  From $\dfrac{9}{11}$ take $\dfrac{4}{5}$ .

$$\dfrac{9}{11} - \dfrac{4}{5} = \dfrac{45}{55} - \dfrac{44}{55} = \dfrac{1}{55}.$$

Ex. 2. From $\dfrac{11}{12}$ of $\dfrac{3}{5}$ take $\dfrac{1}{3}$ of $\dfrac{7}{8}$ .

$$\dfrac{33}{60} - \dfrac{7}{24} = \dfrac{792}{1440} - \dfrac{420}{1440} = \dfrac{372}{1440} = \dfrac{31}{120}.$$

## MULTIPLICATION.

(30.) **Def.** *To multiply one fraction by another, is to take such part or parts of the former as the latter expresses. This is done by multiplying the numerators of the two fractions together for a new numerator, and the denominators for a new denominator.*

Ex. $\dfrac{3}{4} \times \dfrac{5}{7} = \dfrac{15}{28}$; for $\dfrac{3}{4}$ multiplied by $\dfrac{5}{7}$ is, according to the definition of multiplication, $\dfrac{5}{7}$ of $\dfrac{3}{4}$ or $\dfrac{15}{28}$, (Art. 15.)

Compound fractions must be reduced to simple ones, and mixed numbers to improper fractions, and they may then be multiplied as before.

Ex. 1. Multiply $\dfrac{2}{5}$ of $\dfrac{9}{13}$ by $7\dfrac{1}{8}$ .

$\dfrac{2}{5}$ of $\dfrac{9}{13} = \dfrac{18}{65}$ ; and $7\dfrac{1}{8} = \dfrac{57}{8}$ ; therefore their product

is $\dfrac{18}{65} \times \dfrac{57}{8} = \dfrac{1026}{520} = 1\dfrac{506}{520} = 1\dfrac{253}{260}$ .

Ex. 2.

Ex. 2. Multiply $\dfrac{2}{217}$ by 7.

$$\dfrac{2}{217} \times 7 = \dfrac{2 \times 7}{217} = \dfrac{2}{31}.$$

Hence it appears, that a fraction may be multiplied by a whole number, by dividing the denominator by that number, when this division can take place.

## DIVISION.

(31.) *To divide one fraction by another, or to determine how often one is contained in the other, invert the numerator and denominator of the divisor, and proceed as in multiplication.*

Ex. $\dfrac{3}{4}$ divided by $\dfrac{5}{7}$ is $\dfrac{3}{4} \times \dfrac{7}{5} = \dfrac{21}{20} = 1\dfrac{1}{20}.$

For, from the nature of division, the divisor multiplied by the quotient must produce the dividend; therefore $\dfrac{5}{7} \times$ quotient $= \dfrac{3}{4}$; let these equal quantities be multiplied by the same quantity $\dfrac{7}{5}$, and the products must be equal; that is, $\dfrac{7}{5} \times \dfrac{5}{7} \times$ quotient $= \dfrac{3}{4} \times \dfrac{7}{5}$ or $\dfrac{35}{35} \times$ quotient $= \dfrac{21}{20}$; but $\dfrac{35}{35} = 1$ (Art. 14); therefore the quotient $= \dfrac{21}{20}$ according to the rule. And the same method of proof is applicable to all cases.

Compound fractions must be reduced to simple ones, and mixed numbers to improper fractions, before the rule can be applied.

Ex.

Ex. Divide $\dfrac{5}{9}$ of $\dfrac{4}{7}$ by $3\dfrac{1}{3}$.

$\dfrac{5}{9}$ of $\dfrac{4}{7} = \dfrac{20}{63}$, and $3\dfrac{1}{3} = \dfrac{10}{3}$; therefore the quotient is $\dfrac{20}{63} \times \dfrac{3}{10} = \dfrac{60}{630} = \dfrac{2}{21}$.

## ON DECIMAL FRACTIONS.

(32.) In order to lessen the trouble which in many cases attends the use of vulgar fractions, decimal fractions have been introduced, which differ from the former in this respect, that their denominators are always 10 or some power of 10, as 100, 1000, 10000, &c. and instead of *writing* the denominator under the numerator, it is *expressed* by pointing off, from the right of the numerator, as many figures as there are cyphers in the denominator; thus, .2, .23, .127, .0013, 43.7, signify respectively $\dfrac{2}{10}$, $\dfrac{23}{100}$, $\dfrac{127}{1000}$, $\dfrac{13}{10000}$, $43\dfrac{7}{10}$ or $\dfrac{437}{10}$.

(33.) Cor. 1. The value of each figure in a decimal, decreases from the left to the right in a tenfold proportion; that is, each figure is ten times as great as if it were removed one place to the right, as in whole numbers; thus .2, .02, .002, are $\dfrac{2}{10}$, $\dfrac{2}{100}$, $\dfrac{2}{1000}$, &c. and the decimal .127 is one tenth, two hundredths and seven thousandths of an unit.

(34.) Cor.

(34.) Cor. 2. Adding cyphers to the right of a decimal does not alter it's value; thus, .2, .20, .200, or $\dfrac{2}{10}$, $\dfrac{20}{100}$, $\dfrac{200}{1000}$ are equal to each other, the numerator and denominator having been multiplied by the same number. (See Art. 11.)

(35.) Cor. 3. Decimals may be reduced to a common denominator by adding cyphers to the right, where it is necessary, till the number of decimal places is the same in all.

Ex. .5, .01 and .311 reduced to a common denominator, are .500, .010 and .311 ; that is, $\dfrac{500}{1000}$, $\dfrac{10}{1000}$ and $\dfrac{311}{1000}$.

As decimals are only fractions of a particular description, their operations must depend upon the principles already laid down.

## ADDITION OF DECIMALS.

(36.) *To find the sum of any number of decimals, place the figures in such a manner that those of the same denomination may stand under each other; add them together as in whole numbers, and place the decimal point in the sum under the other points.*

Ex. Add together 7.9, 51.43 and .0118.

These, when reduced to a common denominator, are

are 7·9000, 51.4300 and .0118; and proceeding according to the rule,

$$7.9000$$
$$51.4300$$
$$.0118$$

59.3418 is the sum required.   (Art. 26.)

In the operation, the cyphers may be omitted; thus,

$$7·9$$
$$51.43$$
$$.0118$$

$$59.3418$$

## SUBTRACTION.

(37.) *To find the difference of two decimals, place the figures of the same denomination under each other; then subtract as in whole numbers, and place the decimal point under the other points.*

From 61.3 take 42.012.

These, reduced to a common denominator, are 61.300 and 42.012; therefore their difference is 19.288 (Art. 28).   In the operation, the cyphers may be omitted; thus,

$$61.3$$
$$42.012$$

$$19.288$$

MULTI-

## MULTIPLICATION.

(38.) *To multiply one decimal by another, multiply the figures as in whole numbers, and point off as many decimal places in the product as there are in the multiplier and the multiplicand together.*

Ex. $51.3 \times 4.6 = 235.98$. For, $\dfrac{513}{10} \times \dfrac{46}{10} = \dfrac{23598}{100}$

= (according to the decimal notation) 235.98. And a similar proof may be given in all other cases.

(39.) When there are fewer figures in the product than there are decimals in the multiplier and multiplicand together, cyphers must be annexed to the *left* of the product, that the decimal places may be properly represented.

Ex. $.25 \times .3 = .075$; for $\dfrac{25}{100} \times \dfrac{3}{10} = \dfrac{75}{1000} =$ (according to the decimal notation) .075.

## DIVISION.

(40.) *Division in decimals is performed as in whole numbers, observing to point off as many decimals in the quotient as the number of decimal places in the dividend exceeds the number in the divisor.*

Ex. Divide 77.922 by 3.7.

$\dfrac{77.922}{3.7} = 21.06$: here there are three decimals in the dividend, and one in the divisor; therefore, there are two in the quotient.

The truth of this rule is apparent from the nature of multiplication; for, the product of the divisor and

quotient

quotient is the dividend ; there are, therefore, as many places of decimals in the dividend, as there are in the divisor and quotient together (Art. 38); consequently, there are as many in the quotient as the number in the dividend exceeds the number in the divisor.

(41.) If figures be wanting to make up the proper number of decimal places, cyphers must be added to the *left*.

Ex. Divide .336 by 42.

$\frac{336}{42} = 8$ ; and as the quotient of .336 divided by 42 must contain three decimal places, that quotient is .008. For, $\frac{336}{1000}$ divided by 42 is $\frac{336}{42000}$, or $\frac{8}{1000}$ (Art. 9); that is (according to the decimal notation) .008 (Art. 32).

(42.) When the dividend does not contain as many decimals as the divisor, cyphers must be added to the right of the decimals in the dividend, till that is the case.

Ex. Divide 36 by .012.

36=36.000 ; and 36.000 divided by .012 is 3000, according to the rule.


# REDUCTION.

(43.) *To reduce a vulgar fraction to a decimal.*

Add cyphers at pleasure, as decimals, in the numerator, and divide by the denominator according to the rule for the division of decimals. The truth of this rule is evident from Art. 10.

Ex. 1.

Ex. 1. $\frac{3}{4} = \frac{3.00}{4} = .75.$

Ex. 2. $\frac{7}{8} = \frac{7.000}{8} = .875.$

Ex. 3. $\frac{4}{625} = \frac{4.0000}{625} = .0064.$

Ex. 4. $\frac{1}{3} = \frac{1.000 \&c.}{3} = .333 \&c.$

Ex. 5. $\frac{4}{33} = \frac{4.0000 \&c.}{33} = .1212 \&c.$

(44.) In some cases, as in the two last examples, the vulgar fraction cannot exactly be made up of tenths, hundredths, &c. but the decimal will go on without ever coming to an end, the same figure or figures recurring in the same order; but though we cannot represent the exact value of the vulgar fraction, yet, by increasing the number of decimal places, we may approach to it as near as we please. Thus, $\frac{1}{9} = .1111$ &c. now .1, or $\frac{1}{10}$, is less than the true value by $\frac{1}{90}$; .11, or $\frac{11}{100}$, is too little by $\frac{1}{900}$; &c.

Decimals of this kind are called *recurring*, or *circulating* decimals.

(45.) *To find the value of a decimal of one denomination in terms of a lower denomination.*

This may be done by the rule laid down in Art. 22.

Ex.

Ex. Required the value of .615625£.

.615625£.
20
_____

12.312500 *shillings*
12
_____

3.7500    *pence*
4
_____

3.00     *farthings*
_____

The value required is $\overset{s.}{12} : \overset{d.}{3} : \overset{q.}{3}$.

First,    .615625£. = 12.3125 *shillings*.

Next,    .3125s.    = 3.75    *pence.*

Lastly, .75d.     = 3      *farthings.*

(46.) *To reduce a quantity to a decimal of a superior denomination.*

Divide the quantity by the number of integers of it's denomination contained in one of the superior denomination, and the quotient is the decimal required.

Ex. 1. What decimal of a shilling is three-pence?

12)3.00
_____

.25 Answ.
_____

For, in the denomination shillings, it's numerical value must be $\frac{1}{12}$ of it's value in the denomination pence.

Ex. 2. What decimal of a pound is $\overset{s.}{13} : \overset{d.}{4} : \overset{q.}{3}$?

4) 3.00
_____

12) 4.75
_____

20)13.3958333 &c.
_____

.66979166 &c.
_____

First,

First, we find what decimal of a penny $\overset{q.}{3}$ is; this, by the rule, is .75; then, what decimal of a shilling $\overset{d.}{4} : \overset{q.}{3}$ or 4.75$d$. is; this is found in the same manner to be .3958333 &c. lastly, we find, by the same rule, what decimal of a pound 13.3958333 &c. *sh.* is; which appears to be .66979166 &c.

The conclusion will be the same if we reduce the quantity to a vulgar fraction (Art. 23), and this fraction to a decimal (Art. 43).

The proofs of the rules for the management of vulgar and decimal fractions, here given, are necessarily confined to particular instances, but the same reasoning may be applied in every case; and by using general signs, the proofs may be made general.

THE

# THE

# ELEMENTS OF ALGEBRA.

## PART I.

### DEFINITIONS AND EXPLANATION OF SIGNS.

(47.) THE method of representing the relation of abstract quantities by letters and characters, which are made the signs of such quantities and their relations, is called *Algebra*.

Known or determined quantities are usually represented by the first letters of the alphabet, $a, b, c, d$, &c. and unknown or undetermined quantities, by the last, $y, x, w$, &c.

The following signs are made use of to express the relations which the quantities bear to each other.

(48.) + *Plus*, signifies that the quantity to which it is prefixed must be added. Thus, $a + b$ signifies that the quantity represented by $b$ is to be added to the quantity represented by $a$ ; if $a$ represent 5, and $b$, 7, then $a + b$ represents 12.

If no sign be placed before a quantity, the sign + is understood. Thus, $a$ signifies $+ a$. Such quantities are called *positive* quantities.

(49.) − *Minus*, signifies that the quantity to which

it

it is prefixed must be subtracted. Thus, $a-b$ signifies that $b$ must be taken from $a$; if $a$ be 7, and
$b$, 5, $a-b$ expresses 7 diminished by 5, or 2.

Quantities to which the sign — is prefixed are
called *negative* quantities.

(50.) × *Into*, signifies that the quantities between
which it stands are to be multiplied together. Thus,
$a \times b$ signifies that the quantity represented by $a$ is to
be multiplied by the quantity represented by $b$*.

This sign is frequently omitted; thus $abc$ signifies
$a \times b \times c$. Or a full point is used instead of it; thus
$1 \times 2 \times 3$, and $1.2.3$, signify the same thing.

(51.) If in multiplication the same quantity be
repeated any number of times, the product is usually
expressed by placing, above the quantity, the number
which represents how often it is repeated; thus $a$,
$a \times a$, $a \times a \times a$, $a \times a \times a \times a$, and $a^1$, $a^2$, $a^3$, $a^4$,
have respectively the same signification. These
quantities are called *powers*; thus $a^1$, is called the
first power of $a$; $a^2$, the second power, or square of
$a$; $a^3$, the third power, or cube of $a$, &c.

The numbers 1, 2, 3, &c. are called the *indices* of
$a$; or *exponents of the powers* of $a$.

(52.) ÷ *Divided by*, signifies that the former of
the quantities between which it is placed is to be
divided by the latter. Thus, $a \div b$ signifies that the
quantity $a$ is to be divided by $b$.

The division of one quantity by another is frequently represented by placing the dividend over the
divisor

---

* By quantities, we understand such magnitudes as can be
represented by numbers; we may therefore without impropriety
speak of the multiplication, division, &c. of quantities by each
other.

divisor with a line between them, in which case the expression is called a fraction. Thus, $\frac{a}{b}$ signifies $a$ divided by $b$; and $a$ is the numerator, and $b$ the denominator of the fraction; also, $\frac{a+b+c}{e+f+g}$ signifies that $a$, $b$, and $c$ added together, are to be divided by $e$, $f$, and $g$ added together; see Art. 10.

(53.) A quantity in the denominator of a fraction is also expressed by placing it in the numerator, and prefixing the negative sign to it's index; thus, $a^{-1}$, $a^{-2}$, $a^{-3}$, $a^{-n}$, signify $\frac{1}{a^1}$, $\frac{1}{a^2}$, $\frac{1}{a^3}$, $\frac{1}{a^n}$ respectively; these are called the negative powers of $a$.

(54.) The sign $\smallfrown$ between two quantities signifies their *difference*. Thus, $a \smallfrown x$, is $a - x$ or $x - a$, according as $a$ or $x$ is the greater; and $a \mp x$ signifies the sum or difference of $a$ and $x$.

(55.) A line drawn over several quantities signifies that they are to be taken collectively, and it is called a *vinculum*. Thus, $\overline{a-b+c} \times \overline{d-e}$ signifies that the quantity represented by $a - b + c$ is to be multiplied by the quantity represented by $d - e$. Let $a$ stand for 6; $b$, 5; $c$, 4; $d$, 3; and $e$, 1; then $a - b + c$ is $6 - 5 + 4$, or 5; and $d - e$ is $3 - 1$, or 2; therefore $\overline{a - b + c} \times \overline{d - e}$ is $5 \times 2$ or 10. $\overline{ab - cd} \times \overline{ab - cd}$ or $\overline{ab - cd}|^2$ signifies that the quantity represented by $ab - cd$ is to be multiplied by itself.

(56.) $=$ *Equal to*, signifies that the quantities between which it is placed are equal to each other; thus,
$$ax - by$$

$ax - by = cd + ad$, signifies that the quantity $ax - by$ is equal to the quantity $cd + ad$.

(57.) The *square root* of any proposed quantity is that quantity whose square, or second power, gives the proposed quantity. The *cube root*, is that quantity whose cube gives the proposed quantity, &c.

The signs $\sqrt{}$, or $\sqrt[2]{}$, $\sqrt[3]{}$, $\sqrt[4]{}$, &c. are used to express the square, cube, biquadrate, &c. roots of the quantities before which they are placed.

$$\sqrt[2]{a^2} = a, \quad \sqrt[3]{a^3} = a, \quad \sqrt[4]{a^4} = a, \quad \&c.$$

These roots are also represented by the fractions $\frac{1}{2}$, $\frac{1}{3}$, $\frac{1}{4}$, &c., placed a little above the quantities, to the right. Thus, $a^{\frac{1}{2}}$, $a^{\frac{1}{3}}$, $a^{\frac{1}{4}}$, $a^{\frac{1}{n}}$, represent the square, cube, fourth and $n^{th}$ root of $a$, respectively; $a^{\frac{5}{2}}$, $a^{\frac{7}{3}}$, $a^{\frac{3}{5}}$, represent the square root of the fifth power, the cube root of the seventh power, the fifth root of the cube of $a$.

(58.) If these roots cannot be exactly determined, the quantities are called *irrational*, or *surds*.

(59.) Points are made use of to denote proportion thus, $a : b :: c : d$, signifies that $a$ bears the same proportion to $b$ that $c$ bears to $d$.

(60.) The number prefixed to any quantity, and which shews how often it is to be taken, is called it's *coefficient*. Thus, in the quantities $7\,ax$, $6\,by$, $3\,dz$, 7, 6 and 3 are called the coefficients of $ax$, $by$, and $dz$, respectively.

When no number is prefixed, the quantity is to be taken once, or the coefficient 1 is understood.

These

These numbers are sometimes represented by letters, which are called coefficients.

(61.) *Similar,* or *like* algebraical quantities, are such as differ only in their coefficients; $4a$, $6ab$, $9a^2$, $3a^2bc$, are respectively similar to $15a$, $3ab$, $12a^2$. $15a^2bc$, &c.

*Unlike* quantities are different combinations of letters; thus, $ab$, $a^2b$, $ab^2$, $abc$, &c. are unlike.

(62.) A quantity is said to be a *multiple* of another, when it contains it a certain number of times exactly; thus, $16a$ is a multiple of $4a$, as it contains it exactly four times.

(63.) A quantity is called a *measure* of another, when the former is contained in the latter a certain number of times exactly; thus, $4a$ is a measure of $16a$.

(64.) When two numbers have no common measure but unity, they are said to be *prime* to each other.

(65.) A *simple* algebraical quantity is one which consists of a single term, as $a^2bc$.

(66.) A *binomial* is a quantity consisting of two terms, as $a+b$, or $2a-3bx$. A *trinomial* is a quantity consisting of three terms, as $2a+bd+3c$.

The following examples will serve to illustrate the method of representing quantities algebraically.

Let $a=8$, $b=7$, $c=6$, $d=5$ and $e=1$; then,

$3a-2b+4c-e = 24-14+24-1 = 33$.

$ab+ce-bd = 56+6-35 = 27$.

$$\frac{a+b}{c-e} + \frac{3b-2c}{a-d} = \frac{8+7}{6-1} + \frac{21-12}{8-5} = \frac{15}{5} + \frac{9}{3} = 6.$$

$d^2 \times \overline{a-c} - 3ce^2 + d^3 = 25 \times 2 - 18 + 125 = 50 - 18 + 125 = 157$.

AXIOMS.

## AXIOMS.

(67.) If equal quantities be added to equal quantities, the sums will be equal.

(68.) If equal quantities be taken from equal quantities, the remainders will be equal.

(69.) If equal quantities be multiplied by the same, or equal quantities, the products will be equal.

(70.) If equal quantities be divided by the same, or equal quantities, the quotients will be equal.

(71.) If the same quantity be added to and subtracted from another, the value of the latter will not be altered.

(72.) If a quantity be both multiplied and divided by another, it's value will not be altered.

## ADDITION OF ALGEBRAICAL QUANTITIES.

(73.) *The addition of algebraical quantities is performed by connecting those that are* unlike *with their proper signs, and collecting those that are* similar *into one sum.*

Ex. 1. Add together the following *unlike* quantities;

$$ax$$
$$- by$$
$$+ e^2$$
$$- ed$$

Sum $ax - by + e^2 - ed.$

Ex. 2.

## Ex. 2.

$$a + 2b - c$$
$$d - 5e + f$$

Sum $a + 2b - c + d - 5e + f.$

It is immaterial in what order the quantities are set down, if we take care to prefix to each it's proper sign.

When any terms are *similar*, they may be incorporated, and the general expression for the sum shortened.

1ˢᵗ. When *similar* quantities have the *same* sign, their sum is found by taking the sum of the coefficients with that sign, and annexing the common letters.

<table>
<tr><td></td><td>Ex. 4.</td></tr>
<tr><td>Ex. 3.</td><td>$4'a^2c - 10bde$</td></tr>
<tr><td>$5a - 3b$</td><td>$6a^2c - 9bde$</td></tr>
<tr><td>$4a - 7b$</td><td>$11a^2c - 3bde$</td></tr>
<tr><td>Sum $9a - 10b$</td><td>Sum $21a^2c - 22bde$</td></tr>
</table>

The reason is evident; $5a$ to be added, together with $4a$ to be added, makes $9a$ to be added; and $3b$ to be subtracted, together with $7b$ to be subtracted, is $10b$ to be subtracted.

2ᵈ. If *similar* quantities have *different* signs, their sum is found by taking the difference of the coefficients with the sign of the greater, and annexing the common letters as before.

## Ex. 5.

$$7a + 3b$$
$$-5a - 9b$$

Sum $2a - 6b$

In

In the first part of the operation we have 7 times $a$ to add, and 5 times $a$ to take away; therefore upon the whole we have $2a$ to add. In the latter part, we have 3 times $b$ to add, and 9 times $b$ to take away; *i. e.* we have upon the whole 6 times $b$ to take away; and thus the sum of all the quantities is $2a - 6b$.

### Ex. 6.

$$a + b$$
$$a - b$$

Sum $2a$

If several similar quantities are to be added together, some with positive and some with negative signs, take the difference between the sum of the positive, and the sum of the negative coefficients, prefix the sign of the greater sum, and annex the common letters.

### Ex. 7.

$$3a^2 + 4bc - \quad e^2 + 10$$
$$-5a^2 + 6bc + \quad 2e^2 - 15$$
$$-4a^2 - 9bc - 10e^2 + 21$$

Sum $-6a^2 + \quad bc - \quad 9e^2 + 16$

The method of reasoning in this case is the same as in the last example.

### Ex. 8.

$$4ac - 15bd + \quad ex$$
$$11ac + \quad 7b^2 - 19ex$$
$$-41a^2 + \quad 6bd - \quad 7de$$

Sum $15ac - 41a^2 - 9bd + 7b^2 - 18ex - 7de$

### Ex. 9.

## Ex. 9.

$$px^3 - qx^2 - rx$$
$$ax^3 - bx^2 - x$$

Sum $\overline{p+a}.x^3 - \overline{q+b}.x^2 - \overline{r+1}.x$

In this example, the coefficients of $x$ and it's powers are united; $\overline{p+a}.x^3 = px^3 + ax^3$; also — $\overline{q+b}.x^2 = -qx^2 - bx^2$, because the negative sign affects the whole quantity under the vinculum; and $-\overline{r+1}.x = -rx - x$.

## SUBTRACTION.

(74.) *Subtraction, or the taking away of one quantity from another, is performed by changing the sign of the quantity to be subtracted, and then adding it to the other by the rules laid down in Art.* 73.

### Ex. 1.

From $2bx$ take $cy$, and the difference is properly represented by $2bx - cy$; because the $-$ prefixed to $cy$, shews that it is to be subtracted from the other; and $2bx - cy$ is the sum of $2bx$ and $-cy$, Art. 73.

### Ex. 2.

Again, from $2bx$ take $-cy$, and the difference is $2bx + cy$; because $2bx = 2bx + cy - cy$ Art. 71, take away $-cy$ from these equal quantities, and the differences will be equal; *i. e.* the difference between $2bx$ and $-cy$ is $2bx + cy$, the quantity which arises from adding $+cy$ to $2bx$.

Ex. 3.

Ex. 3.

From $\quad a + b$

take $\quad a - b$

Difference $* + 2b$

Ex. 4.

From $\quad 6a - 12b$

take $\quad -5a - 10b$

Diff. $\quad 11a - 2b$

Ex. 5.

From $\quad 5a^2 + 4ab - 6xy$

take $\quad 11a^2 + 6ab - 4xy$

Diff. $-6a^2 - 2ab - 2xy$

Ex. 6.

From $\quad 4a - 3b + 6c - 11$

take $\quad 10x + a - 15 - 2y$

Diff. $-10x + 3a - 3b + 4 + 6c + 2y$

Ex. 7.

From $ax^3 - bx^2 + x$

take $px^3 - qx^2 + rx$

Diff. $\overline{a - p} . x^3 - \overline{b - q} . x^2 + \overline{1 - r} . x$

In this example the coefficients are united; $\overline{a - p} . x^3$ is equal to $ax^3 - px^3$; $-\overline{b - q} . x^2$ is equal to $-bx^2 + qx^2$; and $\overline{1 - r} . x = x - rx$.

## MULTIPLICATION.

(75.) The multiplication of simple algebraical quantities must be represented according to the notation pointed out Art. 50.

c $\qquad$ Thus,

Thus, $a \times b$, or $ab$, represents the product of $a$ multiplied by $b$; $abc$, the product of the three quantities $a$, $b$ and $c$.

It is also indifferent in what order they are placed, $a \times b$ and $b \times a$ being equal.

For, $1 \times a = a \times 1$, or 1 taken $a$ times is the same with $a$ taken once; also, $b$ taken $a$ times, or $b \times a$, is $b$ times as great as 1 taken $a$ times; and $a$ taken $b$ times, or $a \times b$, is $b$ times as great as $a$ taken once; therefore (Art. 69) $b \times a = a \times b$. Also, $abc = cab = bca = acb$, &c. for, as in the former case, $1 \times a \times b = a \times b \times 1$; and $c \times a \times b$ is $c$ times as great as $1 \times a \times b$; also $a \times b \times c$ is $c$ times as great as $a \times b \times 1$; therefore $a \times b \times c = c \times a \times b$ (Art. 69); and a similar proof may be applied to the other cases.

(76.) To determine the *sign* of the product, observe the following rule:

*If the multiplier and multiplicand have the* same *sign, the product is positive; if they have* different *signs, it is negative.*

1$^{st}$. $+a \times +b = +ab$; because in this case $a$ is to be taken positively $b$ times; therefore the product $ab$ must be positive.

2$^{d}$. $-a \times +b = -ab$; because $-a$ is to be taken $b$ times; that is, we must take $-ab$.

3$^{d}$. $+a \times -b = -ab$; for a quantity is said to be multiplied by a negative number $-b$, if it be subtracted $b$ times; and $a$ subtracted $b$ times is $-ab$. This also appears from Art. 79. Ex. 2.

4$^{th}$. $-a \times -b = +ab$. Here $-a$ is to be subtracted $b$ times; that is, $-ab$ is to be subtracted; but subtracting $-ab$ is the same as adding $+ab$ (Art. 74); therefore we have to add $+ab$.

The

The $2^{d}$ and $4^{th}$ cases may be thus proved; $a - a = o$, multiply both sides by $b$, and $ab$ together with $-a \times b$ must be equal to $b \times o$, or nothing; therefore $-a$ multiplied by $b$ must give $-ab$, a quantity which when added to $ab$ makes the sum nothing.

Again, $a - a = o$; multiply both sides by $-b$, then $-ab$ together with $-a \times -b$ must be $=o$; therefore $-a \times -b = +ab$.

(77.) If the quantities to be multiplied have coefficients, these must be multiplied together as in common arithmetic; the sign and the literal product being determined by the preceding rules.

Thus, $3a \times 5b = 15ab$; because $3 \times a \times 5 \times b = 3 \times 5 \times a \times b = 15ab$ (Art. 75); $4x \times -11y = -44xy$; $-9b \times -5c = +45bc$; $-6d \times 4m = -24md$.

(78.) The powers of the same quantity are multiplied together by adding the indices; thus, $a^{2} \times a^{3} = a^{5}$; for $aa \times aaa = aaaaa$. In the same manner, $a^{m} \times a^{n} = a^{m+n}$; and $-3a^{2}x^{3} \times 5axy^{2} = -15a^{3}x^{4}y^{2}$.

(79.) If the multiplier or multiplicand consist of several terms, each term of the latter must be multiplied by every term of the former, and the sum of all the products taken, for the whole product of the two quantities.

Ex. 1.   Mult.  $a + b$
     by     $c + d$
_____
Prod.  $ac + bc + ad + bd$
_____

Here $a + b$ is to be added to itself $c + d$ times, i. e. times and $d$ times.

Ex. 2.

Ex. 2.   Mult. $a+b$
     by     $c-d$

Prod. $ac+bc-ad-bd$

Here $a+b$ is to be taken $c-d$ times; that is, $c$ times wanting $d$ times; or $c$ times positively and $d$ times negatively.

Ex. 3.  Mult. $a+b$          Ex. 4.  Mult. $a+b$
     by      $a+b$                by      $a-b$

$a^2+ab$                                $a^2+ab$
   $+ab+b^2$                              $-ab-b^2$

Prod. $a^2+2ab+b^2$          Prod. $a^2 \quad * -b^2$

Ex. 5.  Mult. $3a^2-5bd$
     by     $-5a^2+4bd$

$-15a^4+25a^2bd$
        $+12a^2bd-20b^2d^2$

Prod. $-15a^4+37a^2bd-20b^2d^2$

Ex. 6.  Mult. $a^2+2ab+b^2$
     by     $a^2-2ab+b^2$

$a^4+2a^3b+a^2b^2$
     $-2a^3b-4a^2b^2-2ab^3$
               $+a^2b^2+2ab^3+b^4$

Prod. $a^4 \quad * \quad -2a^2b^2 \quad * \quad +b^4$

Ex. 7.  Mult. $1-x+x^2-x^3$
     by     $1+x$

$1-x+x^2-x^3$
    $+x-x^2+x^3-x^4$

Prod. $1 \quad * \quad * \quad * \quad -x^4$

Ex. 8.

Ex. 8.  Mult. $x^2 - px + q$
    by    $x + a$

$$x^3 - px^2 + qx$$
$$+ ax^2 - apx + aq$$

Prod. $x^3 - \overline{p-a}.x^2 + \overline{q-ap}.x + aq$

Here the coefficients of $x^2$ and $x$. are collected;
$-\overline{p-a}.x^2 = -px^2 + ax^2$; and $\overline{q-ap}.x = qx - apx$.

## SCHOLIUM.

(80.) The method of determining the sign of a product from the consideration of abstract quantities, has been found fault with by some algebraical writers, who contend that $-a$, without reference to other quantities, is imaginary, and consequently not the object of reason or demonstration. In answer to this objection we may observe, that whenever we make use of the notation $-a$, and say it signifies a quantity to be subtracted, we make a tacit reference to other quantities.

Thus, in numbers, $-a$ represents a number to be subtracted from those with which it is connected; and when we suppose $-a$ to be taken $b$ times, we must understand that $a$ is to be taken $b$ times from some other numbers. In estimating lines, or distances, $-a$ represents a line, or distance, in a particular direction. The negative sign does not render quantities imaginary, or impossible, but points out the relation of real quantities to others with which they are concerned.

DIVISION.

## DIVISION.

(81.) *To divide one quantity by another, is to determine how often the latter is contained in the former, or what quantity multiplied by the latter will produce the former.*

Thus, to divide $ab$ by $a$ is to determine how often $a$ must be taken to make up $ab$; that is, what quantity multiplied by $a$ will give $ab$; which we know is $b$. From this consideration are derived all the rules for the division of algebraical quantities.

(82.) If the divisor and dividend be affected with *like* signs, the sign of the quotient is $+$ : but if their signs be *unlike*, the sign of the quotient is $-$ .

If $-ab$ be divided by $-a$, the quotient is $+b$; because $-a \times +b$ gives $-ab$; and a similar proof may be given in the other cases.

(83.) In the division of simple quantities, if the coefficient and literal product of the divisor be found in the dividend, the other part of the dividend, with the sign determined by the last rule, is the quotient.

Thus, $\dfrac{abc}{ab} = c$; because $ab$ multiplied by $c$ gives $abc$.

If we first divide by $a$, and then by $b$, the result will be the same; for $\dfrac{abc}{a} = bc$, and $\dfrac{bc}{b} = c$, as before.

(84.) Cor. Hence, any power of a quantity is divided by any other power of the same quantity, by taking the index of the divisor from the index of the dividend.

Thus, $\dfrac{a^5}{a^3} = a^2$; $\dfrac{a^3}{a^6} = \dfrac{1}{a^3} = a^{-3}$ (Art. 53); $\dfrac{a^m}{a^n} = a^{m-n}$.

(85.) If

(85.) If only a part of the product which forms the divisor, be contained in the dividend, the division must be represented according to the direction in Art. 52, and the quantities contained both in the divisor and dividend expunged.

Thus, $15\,a^3b^2c$ divided by $-\underline{3}a^2bx$, or $\dfrac{15\,a^3b^2c}{-3a^2bx} = \dfrac{-5\,abc}{x}$.

First, divide by $-3\,a^2b$, and the quotient is $-5\,abc$; this quantity is still to be divided by $x$ (Art. 83), and as $x$ is not contained in it, the division can only be represented in the usual way; that is, $\dfrac{-5\,abc}{x}$ is quotient.

(86.) If the dividend consist of several terms, and the divisor be a simple quantity, every term of the dividend must be divided by it.

Thus, $\dfrac{a^3x^2 - 5\,abx^3 + 6\,ax^4}{ax^2} = a^2 - 5\,bx + 6\,x^2.$

(87.) When the divisor also consists of several terms, arrange both the divisor and dividend according to the powers of some one letter contained in them; then, find how often the first term of the divisor is contained in the first term of the dividend, and write down this quantity for the first term in the quotient; multiply the whole divisor by it, subtract the product from the dividend, and bring down to the remainder as many other terms of the dividend as the case may require, and repeat the operation till all the terms are brought down.

### Ex. 1.

If $a^2 - 2ab + b^2$ be divided by $a-b$, the operation will be as follows :

$$a-b$$

$$a - b)a^2 - 2ab + b^2(a - b$$
$$a^2 - ab$$

$$-ab + b^2$$
$$-ab + b^2$$

\* \*

The reason of this, and the foregoing rule, is, that as the whole dividend is made up of all it's parts, the divisor is contained in the whole, as often as it is contained in all the parts. In the preceding operation we inquire first, how often $a$ is contained in $a^2$, which gives $a$ for the first term of the quotient, then multiplying the whole divisor by it, we have $a^2 - ab$ to be subtracted from the dividend, and the remainder is $-ab + b^2$, with which we are to proceed as before.

The whole quantity $a^2 - 2ab + b^2$, is in reality divided into two parts by the process, each of which is divided by $a - b$; therefore the true quotient is obtained.

### Ex. 2.

$$a + b)ac + ad + bc + bd(c + d$$
$$ac + bc$$

$$ad + bd$$
$$ad + bd$$

\* \*

Ex. 3.

## Ex. 3.

$$1-x)1 \qquad (1+x+x^2+x^3+\&c.+\dfrac{\text{Remainder}}{1-x}$$

$$\underline{1-x}$$

$$\underline{\begin{array}{l}+x, \\ +x-x^2\end{array}}$$

$$\underline{\begin{array}{l}+x^2 \\ +x^2-x^3\end{array}}$$

$$\underline{\begin{array}{l}+x^3 \\ +x^3-x^4\end{array}}$$

$$+x^4 \ \&c.,$$

## Ex. 4.

$$y-1)y^3-1(y^2+y+1$$
$$\underline{y^3-y^2}$$
$$\underline{\begin{array}{l}+y^2 \\ +y^2-y\end{array}}$$
$$\underline{\begin{array}{l}+y-1 \\ +y-1\end{array}}$$
$$*$$

## Ex. 5.

$$x-y)x^m-y^m(x^{m-1}+x^{m-2}y+x^{m-3}y^2\ldots\ldots+y^{m-1}$$
$$\underline{x^m-x^{m-1}y}$$
$$\underline{\begin{array}{l}+x^{m-1}y \\ +x^{m-1}y-x^{m-2}y^2\end{array}}$$
$$\underline{\begin{array}{l}+x^{m-2}y^2 \\ +x^{m-2}y^2-x^{m-3}y^3\end{array}}$$
$$+x^{m-3}y^3 \ \&c.$$

## Ex. 6.

## Ex. 6.

$$x-a)x^3-px^2+qx-r(x^2+\overline{a-p}\,.\,x+a^2-pa+q$$
$$x^3-ax^2$$

$$\overline{a-p}\,.\,x^2+qx$$
$$\overline{a-p}\,.\,x^2-\overline{a^2-pa}\,.\,x$$

$$+\overline{a^2-pa+q}\,.\,x-r$$
$$\overline{a^2-pa+q}\,.\,x-\overline{a^3-pa^2+qa}$$

Remainder   $a^3-pa^2+qa-r$

## ON THE TRANSFORMATION OF FRACTIONS TO OTHERS OF EQUAL VALUE.

(88.) If the signs of all the terms both in the numerator and denominator of a fraction be changed, it's value will not be altered. For $\dfrac{-ab}{-a}=+b=\dfrac{+ab}{+a}$; and $\dfrac{ab}{-a\text{-}}=-b=\dfrac{-ab}{a}$.

(89.) If the numerator and denominator of a fraction be both multiplied, or both divided by the same quantity, it's value is not altered.

For $\dfrac{ac}{bc}=\dfrac{a}{b}$ (Art. 85).

Hence, a fraction is reduced to it's lowest terms, by dividing both the numerator and denominator by the greatest quantity that measures them both.

(90.) *The greatest common measure of two quantities is found by arranging them according to the powers of some letter, and then dividing the greater by the less, and the preceding divisor always by the last remainder, till the remainder is nothing; the last divisor is the greatest common measure required.*

Let

Let $a$ and $b$ be the two quantities, and let $b$ be contained in $a$, $p$ times, with a remainder $c$; again, let $c$ be contained in $b$, $q$ times with a remainder $d$, and so on, till nothing remains; let $d$ be the last divisor, and it will be the greatest common measure of $a$ and $b$.

$$b)a(p$$
$$\text{---}$$
$$c)b(q$$
$$\text{---}$$
$$d)c(r$$
$$\text{---}$$
$$o$$

(91.) The truth of this rule depends upon these two principles;-

1$^{st}$. If one quantity measure another, it will also measure any multiple of that quantity. Let $x$ measure $y$ by the units in $n$, then it will measure $cy$ by the units in $nc$.

2$^{d}$. If a quantity measure two others, it will measure their sum or difference. Let $a$ be contained in $x$, $m$ times, and in $y$, $n$ times; then $ma = x$ and $na = y$; therefore $x \pm y = ma \pm na = \overline{m \pm n}.a$; i. e. $a$ is contained in $x \pm y$, $m \pm n$ times, or it measures $x \pm y$ by the units in $m \pm n$.

(92.) Now it appears from the operation (Art. 90), that $a - pb = c$, and $b - qc = d$; every quantity therefore which measures $a$ and $b$, measures $pb$, and $a - pb$, or $c$; hence also it measures $qc$, and $b - qc$, or $d$; that is, every common measure of $a$ and $b$ measures $d$.

It appears also from the division, that $a = pb + c$, $b = qc + d$, $c = rd$; therefore $d$ measures $c$, and $qc$, and $qc + d$ or $b$; hence it measures $pb$, and $pb + c$, or $a$. Every common measure then of $a$ and $b$ measures $d$, and $d$ measures $a$ and $b$; therefore $d$ is their greatest common measure.

### Ex.

To find the greatest common measure of $a^4 - x^4$ and $a^3 - a^2x - ax^2 + x^3$, and to reduce $\dfrac{a^4 - x^4}{a^3 - a^2x - ax^3 + x^3}$ to it's lowest terms.

$$a^3 - a^2x$$

$$a^3 - a^2x - ax^2 + x^3)a^4 - x^4(a + x$$
$$a^4 - a^3x - a^2x^2 + ax^3$$

$$\overline{\qquad\qquad\qquad}$$

$$a^3x + a^2x^2 - ax^3 - x^4$$
$$a^3x - a^2x^2 - ax^3 + x^4$$

$$\overline{\qquad\qquad\qquad}$$

$$2a^2x^2 - 2x^4$$

leaving out $2x^2$, which is found in each term of the remainder, the next divisor is $a^2 - x^2$.

$$a^2 - x^2)a^3 - a^2x - ax^2 + x^3(a - x$$
$$a^3 - ax^2$$

$$\overline{\qquad\qquad}$$

$$-a^2x + x^3$$
$$-a^2x + x^3$$

$$\overline{\qquad\qquad}$$
$$*$$

$a^2 - x^2$ is therefore the greatest common measure of the two quantities, and if they be respectively divided by it, the fraction is reduced to $\dfrac{a^2 + x^2}{a - x}$, it's lowest terms.

The quantity $2x^2$, found in every term of one of the divisors, $2a^2x^2 - 2x^4$, but not in every term of the dividend, $a^3 - a^2x - ax^2 + x^3$, must be left out; otherwise the quotient will be fractional, which is contrary to the supposition made in the proof of the rule; and by omitting this part, $2x^2$, no common measure of the divisor and dividend is left out; because, by the supposition, no part of $2x^2$ is found in all the terms of the dividend.

(93.) To find the greatest common measure of three quantities, $a$, $b$, $c$; take $d$ the greatest common measure of $a$ and $b$; and the greatest measure of $d$ and $c$, is the greatest common measure required.

Because

Because every common measure of $a$, $b$ and $c$, measures $d$ and $c$; and every measure of $d$ and $c$ measures $a$, $b$ and $c$ (Art. 92); therefore the greatest common measure of $d$ and $c$ must be the greatest common measure of $a$, $b$ and $c$.

(94.) In the same manner, the greatest common measure of four or more quantities may be found.

The greatest common measure of four quantities, $a$, $b$, $c$, $d$, may also be found by taking $x$ the greatest common measure of $a$ and $b$, and $y$ the greatest common measure of $c$ and $d$; then the greatest common measure of $x$ and $y$ will be the common measure required.

(95.) If one number be divided by another, and the preceding divisor by the remainder, according to Art. 90, the remainder will at length be less than any quantity that can be assigned.

For $a = pb + c$; and $b$, and consequently $pb$, is greater than $c$; therefore $pb + c$, or $a$, is greater than $2c$, and $\dfrac{a}{2}$ is greater than $c$; therefore from $a$, a quantity greater than it's half has been taken; in the same manner, when $c$ is the dividend, more than it's half is taken away, and so on: but if from any quantity there be taken more than it's half, and from the remainder more than it's half, and so on, there will, at length, remain a quantity less than any that can be assigned (Euc. 1. x).

(96.) *Fractions are changed to others of equal value with a common denominator, by multiplying each numerator by every denominator except it's own, for the new numerator; and all the denominators together for the common denominator.*

Let

Let $\frac{a}{b}, \frac{c}{d}, \frac{e}{f}$ be the proposed fractions; then $\frac{adf}{bdf}$, $\frac{cbf}{bdf}, \frac{edb}{bdf}$, are fractions of the same value with the former, having the common denominator $bdf$. For $\frac{adf}{bdf} = \frac{a}{b}$; $\frac{cbf}{bdf} = \frac{c}{d}$; and $\frac{edb}{bdf} = \frac{e}{f}$ (Art. 89); the numerator and denominator of each fraction having been multiplied by the same quantity, *viz.* the product of the denominators of all the other fractions.

(97.) When the denominators of the proposed fractions are not prime to each other, find their greatest common measure; multiply both the numerator and denominator of each fraction, by the denominators of all the rest, divided respectively by their greatest common measure; and the fractions will be reduced to a common denominator in lower terms * than they would have been by proceeding according to the former rule.

Thus, $\frac{a}{mx}, \frac{b}{my}, \frac{c}{mz}$ reduced to a common denominator, are $\frac{ayz}{mxyz}$; $\frac{bxz}{mxyz}$; $\frac{cxy}{mxyz}$.

## ON THE ADDITION AND SUBTRACTION OF FRACTIONS.

(98.) *If the fractions to be added have a common denominator, their sum is found by adding the numerators together and retaining the common denominator.*

Thus,

---

* To obtain them in the *lowest* terms, each must be reduced to another of equal value, with the denominator which is the least common multiple of all the denominators. See Art. 374.

Thus, $\dfrac{a}{b} + \dfrac{c}{b} = \dfrac{a+c}{b}$. This follows from the principle laid down in Art. 87.

(99.) If the fractions have not a common denominator they must be transformed to others of the same value, which have a common denominator (Art. 96), and then the addition may take place as before.

Ex. 2.

$$\frac{a}{b} + \frac{c}{d} = \frac{ad}{bd} + \frac{bc}{bd} = \frac{ad + bc}{bd}.$$

Ex. 3.

$$\frac{1}{a+b} + \frac{1}{a-b} = \frac{a-b}{a^2-b^2} + \frac{a+b}{a^2-b^2} = \frac{a-b+a+b}{a^2-b^2} = \frac{2a}{a^2-b^2}.$$

Ex. 4.

$$a + \frac{e}{f} = \frac{af}{f} + \frac{e}{f} = \frac{af+e}{f}.$$ Here $a$ is considered as a fraction whose denominator is unity.

Ex. 5.

$$2 + \frac{a+b}{a-b} + \frac{a-b}{a+b} = \frac{2a^2 - 2b^2}{a^2-b^2} + \frac{a^2 + 2ab + b^2}{a^2-b^2} + \frac{a^2 - 2ab + b^2}{a^2-b^2} = \frac{2a^2 - 2b^2 + a^2 + 2ab + b^2 + a^2 - 2ab + b^2}{a^2-b^2}$$

$$= \frac{4a^2}{a^2-b^2}.$$

(100.) *If two fractions have a common denominator, their difference is found by taking the difference of the numerators and retaining the common denominator.*

Thus, $\dfrac{a}{b} - \dfrac{c}{b} = \dfrac{a-c}{b}$ (See Art. 87).

(101.) If

(101.) If they have not a common denominator, they must be transformed to others of the same value, which have a common denominator, and then the subtraction may take place as before.

**Ex. 2.**

$$\frac{a}{b} - \frac{c}{d} = \frac{ad}{bd} - \frac{bc}{bd} = \frac{ad - bc}{bd}.$$

**Ex. 3.**

$$a - \frac{cd}{b} = \frac{ab}{b} - \frac{cd}{b} = \frac{ab - cd}{b}.$$

**Ex. 4.**

$$\frac{a}{b} - \frac{c+d}{c-d} = \frac{ac - ad}{bc - bd} - \frac{bc + bd}{bc - bd} = \frac{ac - ad - bc - bd}{bc - bd}$$

The sign of $bd$ is negative, because every part of the latter fraction is to be taken from the former.

**Ex. 5.**

$$\frac{a+b}{a-b} - \frac{a-b}{a+b} = \frac{a^2 + 2ab + b^2}{a^2 - b^2} - \frac{a^2 - 2ab + b^2}{a^2 - b^2} =$$

$$\frac{a^2 + 2ab + b^2 - a^2 + 2ab - b^2}{a^2 - b^2} = \frac{4ab}{a^2 - b^2}.$$

## ON THE MULTIPLICATION AND DIVISION OF FRACTIONS.

(102.) *To multiply a fraction by any quantity, multiply the numerator by that quantity and retain the denominator.*

Thus, $\frac{a}{b} \times c = \frac{ac}{b}$. For if the quantity to be divided be $c$ times as great as before, and the divisor the same, the quotient must be $c$ times as great.

(103.) Cor.

(103.) Cor. 1. $\frac{a}{b} \times b = \frac{ab}{b} = a.$ That is, if a fraction be multiplied by it's denominator, the product is the numerator.

(104.) Cor. 2. The result is the same, whether the numerator be multiplied by a given quantity, or the denominator divided by it. Let the fraction be $\frac{ad}{bc}$, and let it's numerator be multiplied by $c$, the result is $\frac{adc}{bc}$, or $\frac{ad}{b}$ (Art. 89), the quantity which arises from the division of it's denominator by $c$.

(105.) *The product of two fractions is found by multiplying the numerators together for a new numerator, and the denominators for a new denominator.*

Let $\frac{a}{b}$ and $\frac{c}{d}$ be the two fractions; then $\frac{a}{b} \times \frac{c}{d} = \frac{ac}{bd}.$ For if $\frac{a}{b} = x$, and $\frac{c}{d} = y$, by multiplying the equal quantities $\frac{a}{b}$ and $x$, by $b$, $a = bx$ (Art. 69); in the same manner, $c = dy$; therefore, by the same axiom, $ac = bdxy$; dividing these equal quantities, $ac$ and $bdxy$, by $bd$, we have $\frac{ac}{bd} = xy = \frac{a}{b} \times \frac{c}{d}.$ (See Art. 70).

(106.) *To divide a fraction by any quantity, multiply the denominator by that quantity, and retain the numerator.*

The fraction $\frac{a}{b}$ divided by $c$, is $\frac{a}{bc}$. Because $\frac{a}{b} = \frac{ac}{bc}$,

D                                                              and

and a $c^{th}$ part of this is $\frac{a}{bc}$; the quantity to be divided being a $c^{th}$ part of what it was before, and the divisor the same.

(107.) COR. The result is the same, whether the denominator is multiplied by the quantity, or the numerator divided by it.

Let the fraction be $\frac{ac}{bd}$; if the denominator be multiplied by $c$, it becomes $\frac{ac}{bdc}$ or $\frac{a}{bd}$; the quantity which arises from the division of the numerator by $c$.

(108.) *To divide one fraction by another, invert the numerator and denominator of the divisor, and proceed as in multiplication.*

Let $\frac{a}{b}$ and $\frac{c}{d}$ be the two fractions, then $\frac{a}{b} \div \frac{c}{d} =$
$\frac{a}{b} \times \frac{d}{c} = \frac{ad}{bc}$.

For if $\frac{a}{b} = x$, and $\frac{c}{d} = y$, then as in Art. 105, $a = bx$, and $c = dy$; also, $ad = bdx$, and $bc = bdy$; therefore by Art. 70, $\frac{ad}{bc} = \frac{bdx}{bdy} = \frac{x}{y} = \frac{a}{b} \div \frac{c}{d}$.

(109.) The rule for multiplying the powers of the same quantity (Art. 78), will hold when one or both of the indices are negative.

Thus, $a^m \times a^{-n} = a^{m-n}$; for $a^m \times a^{-n} = a^m \times \frac{1}{a^n}$ (Art. 53)
$= \frac{a^m}{a^n} = a^{m-n}$; in the same manner, $x^3 \times x^{-5} = \frac{x^3}{x^5} = \frac{1}{x^2}$
$= x^{-2}$.

Again,

Again, $a^{-m} \times a^{-n} = a^{-\overline{m+n}}$; because $a^{-m} \times a^{-n} = \dfrac{1}{a^m}$

$\times \dfrac{1}{a^n}$ (Art. 53), $= \dfrac{1}{a^{m+n}} = a^{-\overline{m+n}}$.

(110.) Cor. If $m = n$, $a^m \times a^{-m} = a^{m-m} = a^o$; also,

$a^m \times a^{-m} = \dfrac{a^m}{a^m} = 1$; therefore $a^o = 1$; according to the notation adopted (Arts. 51. 53).

(111.) The rule for dividing any power of a quantity by any other power of the same quantity (Art. 84) holds, whether those powers are positive or negative.

Thus, $a^m \div a^{-n} = a^m \div \dfrac{1}{a^n}$ (Art. 53), $= a^m \times a^n = a^{m+n}$.

Again, $a^{-m} \div a^{-n} = \dfrac{1}{a^m} \div \dfrac{1}{a^n} = \dfrac{a^n}{a^m}$ (Art. 108.) $= a^{n-m}$

(Art. 84).

(112.) Cor. Hence it appears, that a quantity may be transferred from the numerator of a fraction to the denominator, and the contrary, by changing the sign of it's index. Thus, $\dfrac{a^m \times a^n}{b^p} = \dfrac{a^m}{b^p a^{-n}}$; and $\dfrac{a^m}{a^n b^p}$

$= \dfrac{a^m \times a^{-n}}{b^p}$.

## ON INVOLUTION AND EVOLUTION.

(113.) If a quantity be continually multiplied by itself, it is said to be involved, or raised; and the power to which it is raised, is expressed by the number of times the quantity has been employed in the multiplication.

D 2                    Thus,

Thus, $a \times a$, or $a^2$, is called the second power of $a$; $a \times a \times a$, or $a^3$, the third power; $a \times a \ldots (n)$, or $a^n$, the $n^{th}$ power.

(114.) If the quantity to be involved be negative, the signs of the even powers will be positive, and the signs of the odd powers negative.

For $-a \times -a = a^2$; $-a \times -a \times -a = -a^3$, &c.

(115.) A simple quantity is raised to any power, by multiplying the index of every factor in the quantity by the exponent of the power, and prefixing the proper sign determined by the last article.

Thus, $a^m$ raised to the $n^{th}$ power is $a^{mn}$. Because $a^m \times a^m \times a^m \ldots$ to $n$ factors, by the rule of multiplication, is $a^{mn}$; also, $\overline{ab}|^n = ab \times ab \times ab \times$ &c. to $n$ factors, or $a \times a \times a \ldots$ to $n$ factors $\times b \times b \times b \ldots$ to $n$ factors (Art. 75), $= a^n \times b^n$; and $a^2 b^3 c$ raised to the fifth power is $a^{10} b^{15} c^5$. Also, $-a^m$ raised to the $n^{th}$ power is $\pm a^{mn}$; where the positive or negative sign is to be prefixed, according as $n$ is an even or odd number.

(116.) If the quantity to be involved be a fraction, both the numerator and denominator must be raised to the proposed power (Art. 105).

(117.) If the quantity proposed be a compound one, the involution may either be represented by the proper index, or it may actually take place.

Let

Let $a+b$ be the quantity to be raised to any power.

$$a+b$$
$$a+b$$

$$a^2 + ab$$
$$+ ab + b^2$$

$\overline{a+b}|^2$ or $a^2 + 2ab + b^2$ the square, or $2^d$ power.

$$a+b$$

$$a^3 + 2a^2b + ab^2$$
$$+ a^2b + 2ab^2 + b^3$$

$\overline{a+b}|^3$ or $a^3 + 3a^2b + 3ab^2 + b^3$ the $3^d$ power.

$$a+b$$

$$a^4 + 3a^3b + 3a^2b^2 + ab^3$$
$$+ a^3b + 3a^2b^2 + 3ab^3 + b^4$$

$\overline{a+b}|^4$ or $a^4 + 4a^3b + 6a^2b^2 + 4ab^3 + b^4$ the $4^{th}$ power.

If $b$ be negative, or the quantity to be involved be $a-b$, wherever an odd power of $b$ enters, the sign of the term must be negative (Art. 114).

Hence, $\overline{a-b}|^4 = a^4 - 4a^3b + 6a^2b^2 - 4ab^3 + b^4$.

(118.) *Evolution,* or the extraction of roots, is the method of determining a quantity which raised to a proposed power will produce a given quantity.

(119.) Since the $n^{th}$ power of $a^m$ is $a^{mn}$, the $n^{th}$ root of $a^{mn}$ must be $a^m$; *i. e.* to extract any root of a single quantity, we must divide the index of that quantity by the index of the root required.

(120.) When the index of the quantity is not exactly divisible by the number which expresses the root to be extracted, that root must be represented

according

according to the notation pointed out in Art. 57. Thus, the square, cube, fourth, $n^{th}$ root of $a^2 + x^2$, are respectively represented by $\overline{a^2 + x^2}|^{\frac{1}{2}}$, $\overline{a^2 + x^2}|^{\frac{1}{3}}$, $\overline{a^2 + x^2}|^{\frac{1}{4}}$, $\overline{a^2 + x^2}|^{\frac{1}{n}}$; the same roots of $\dfrac{1}{a^2 + x^2}$, or $\overline{a^2 + x^2}|^{-1}$, are represented by $\overline{a^2 + x^2}|^{-\frac{1}{2}}$, $\overline{a^2 + x^2}|^{-\frac{1}{3}}$, $\overline{a^2 + x^2}|^{-\frac{1}{4}}$, $\overline{a^2 + x}|^{-\frac{n}{r}}$.

(121.) If the root to be extracted be expressed by an odd number, the sign of the root will be the same with the sign of the proposed quantity, as appears by Art. 114.

(122.) If the root to be extracted be expressed by an even number, and the quantity proposed be positive, the root may be either positive or negative. Because either a positive or negative quantity, raised to such a power, is positive (Art. 114).

(123.) If the root proposed to be extracted be expressed by an even number, and the sign of the proposed quantity be negative, the root cannot be extracted; because no quantity, raised to an even power, can produce a negative result. Such roots are called *impossible*.

(124.) Any root of a product may be found by taking that root of each factor, and multiplying the roots, so taken, together.

Thus, $\overline{ab}|^{\frac{1}{n}} = a^{\frac{1}{n}} \times b^{\frac{1}{n}}$; because each of these quantities, raised to the $n^{th}$ power, is $ab$ (Art. 115).

Cor. If $a = b$, then $a^{\frac{1}{n}} \times a^{\frac{1}{n}} = a^{\frac{2}{n}}$; and in the same manner, $a^{\frac{r}{n}} \times a^{\frac{s}{n}} = a^{\frac{r+s}{n}}$.

(125.) Any

(125.) Any root of a fraction may be found by taking that root of both the numerator and denominator (Art. 116).

Thus, the cube root of $\dfrac{a^2}{b^2}$ is $\dfrac{a^{\frac{2}{3}}}{b^{\frac{2}{3}}}$, or $a^{\frac{2}{3}} \times b^{-\frac{2}{3}}$;

and $\overline{\dfrac{a}{b}}\Big|^{\frac{1}{n}} = \dfrac{a^{\frac{1}{n}}}{b^{\frac{1}{n}}}$, or, $a^{\frac{1}{n}} \times b^{-\frac{1}{n}}$.

(126.) *To extract the square root of a* compound *quantity.*

Since the square root of $a^2 + 2ab + b^2$ is $a + b$ (Art. 117), whatever be the values of $a$ and $b$, we may obtain a general rule for the extraction of the square root, by observing in what manner $a$ and $b$ may be derived from $a^2 + 2ab + b^2$.

Having arranged the terms according to the dimensions of one letter, $a$, the square root of the first term, $a^2$, is $a$, the first factor in the root; subtract it's square from the whole

$$a^2 + 2ab + b^2(a + b$$
$$a^2$$
$$\overline{\phantom{aaaaaaa}}$$
$$2a + b)\,2ab + b^2$$
$$2ab + b^2$$
$$\overline{\phantom{aaaaaaa}}$$
$$*\quad *$$

quantity, and bring down the remainder $2ab + b^2$; divide $2ab$ by $2a$, and the result is $b$, the other factor in the root; then multiply the sum of twice the first factor and the second $(2a + b)$, by the second $(b)$, and subtract this product $(2ab + b^2)$ from the remainder. If there be more terms, consider $a + b$ as a new value of $a$; and it's square, that is $a^2 + 2ab + b^2$, having, by the first part of the process, been subtracted from the proposed quantity, divide the remainder by the double of this new value of $a$, for a new factor in the root; and for a new subtrahend, multiply this factor by

twice

twice the sum of the former factors increased by this factor. The process must be repeated till the root, or the necessary approximation to the root, is obtained.

## Ex. 1.

To extract the square root of $a^2 + 2ab + b^2 + 2ac + 2bc + c^2$; or of it's equal $a^2 + \overline{2a+b}.b + \overline{2a+2b+c}.c$.

$$a^2 + \overline{2a+b}.b + \overline{2a+2b+c}.c\,(a+b+c$$
$$\underline{a^2}$$
$$2a+b)\overline{2a+b}.b$$
$$\underline{\overline{2a+b}.b}$$
$$2a+2b+c)\overline{2a+2b+c}.c$$
$$\underline{\overline{2a+2b+c}.c}$$
$$*\qquad*$$

## Ex. 2.

To extract the square root of $a^2 - ax + \dfrac{x^2}{4}$.

$$a^2 - ax + \frac{x^2}{4}\left(a - \frac{x}{2}\right)$$
$$\underline{a^2}$$
$$2a - \frac{x}{2}\Big) - ax + \frac{x^2}{4}$$
$$\underline{-ax + \frac{x^2}{4}}$$
$$*\qquad*$$

Ex. 3.

## Ex. 3.

To extract the square root of $1 + x$.

$$1 + x \left(1 + \frac{x}{2} - \frac{x^2}{8} + \&c.\right)$$

$$1$$

——————

$$2 + \frac{x}{2}\Big) \qquad x$$

$$x . + \frac{x^2}{4}$$

——————

$$2 + x - \frac{x^2}{8}\Big) \qquad - \frac{x^2}{4}$$

$$- \frac{x^2}{4} - \frac{x^3}{8} + \frac{x^4}{64}$$

——————

$$\frac{x^3}{8} - \frac{x^4}{64} \ \&c.$$

——————

(127.) It appears from the second example, that a trinomial $a^2 - ax + \frac{x^2}{4}$, in which four times the product of the first and last terms, is equal to the square of the middle term, is a complete square.

(128.) The method of extracting the cube root is discovered in the same manner.

The cube root of $a^3 + 3a^2b + 3ab^2 + b^3$ is $a + b$ (Arts. 117, 118); and to obtain $a + b$ from this compound quantity, arrange the terms as before, and the cube root of the first term, $a^3$, is $a$ the first factor in the root;

$$a^3 + 3a^2b + 3ab^2 + b^3 \ (a + b$$
$$a^3$$
$$3a^2\overline{) \quad 3a^2b + 3ab^2 + b^3}$$
$$3a^2b + 3ab^2 + b^3$$

——————

$$*$$

subtract

subtract it's cube from the whole quantity, and divide the first term of the remainder by $3a^2$, the result is $b$, the second factor in the root; then subtract $3a^2b + 3ab^2 + b^3$ from the remainder, and the whole cube of $a + b$ has been subtracted. If any quantity be left, proceed with $a + b$ as a new $a$, and divide the last remainder by $3.\overline{a+b}\rvert^2$ for a third factor in the root; and thus any number of factors may be obtained.

## SCHOLIUM.

(129.) The rules above laid down, for the extraction of the roots of compound quantities, are but little used in algebraical or fluxional operations; but it was necessary to give them at full length, for the purpose of investigating rules for the extraction of the square and cube roots in numbers.

The square root of 100 is 10, of 10000 is 100, of 1000000 is 1000, &c. from which consideration it follows, that the square root of a number less than 100 must consist of only one figure, of a number between 100 and 10000 of two places of figures, of any number from 10000 to 1000000, of three places of figures, &c. If then a point be made over every second figure in any number, beginning with the units, the number of points will shew the number of figures, or places, in the square root. Thus the square root of $4\overset{.}{3}5\overset{.}{7}$ consists of two figures, the square root of $5\overset{.}{6}4\overset{.}{7}8$, of three figures, &c.

Let

Let the square root of 4357 be required.

Having pointed it ac-
cording to the direction,
it appears that the root
consists of two places of
figures; let $a + b$ be the
root, where $a$ is the value
of the figure in the ten's

$$4\overset{\cdot}{3}5\overset{\cdot}{7}(60+6 \text{ or } 66$$
$$3600 \quad \text{[the root.}$$

$$120+6) \ 757$$
$$\text{or } 126] \ 756$$

$$1 \text{ remainder.}$$

place, and $b$, of that in the unit's; then is $a$ the nearest
square root of 4300 which does not exceed the true
root, this appears to be 60; subtract the square of 60
($a^2$) from the given number, and the remainder is 757;
divide this remainder by 120 (2$a$), and the quotient
is 6 (the value of $b$,) and the subtrahend, or quantity,
to be taken from the last remainder 757, is 126 × 6,
($\overline{2a+b}.b$) or 756.

It is said that $a$ must be the greatest number whose
square does not exceed 4300: it evidently cannot be
a greater number than this; and if possible let it be
some quantity $x$, less than this; then since $x$ is in the
ten's place and $b$ in the unit's, $x + b$ is less than $a$;
therefore the square of $x + b$, whatever be the value of
$b$, must be less than $a^2$, and consequently $x + b$ less
than the true root.

If the root consist of three places of figures, let $a$
represent the hundreds, and $b$ the tens; then having
obtained $a$ and $b$ as before, let the new value of $a$ be
the hundreds and tens together, and find a new
value of $b$ for the units: and thus the process may
be continued when there are more places of figures in
the root.

(130.) The

(130.) The cyphers being omitted for the sake of expedition, the following rule is obtained from the foregoing process.

Point every second figure beginning with the unit's place, dividing by this process the whole number into seve- ral periods ; find the greatest number whose square is con- tained in the first period,

$$4357(66$$
$$36$$
$$\overline{\phantom{000}}$$
$$126)\ 757$$
$$756$$
$$\overline{\phantom{000}}$$
$$1$$

this is the first figure in the root ; subtract it's square from the first period, and to the remainder bring down the next period ; divide this quantity, omitting the last figure, by twice the part of the root already obtained, and annex the result to the root and also to the divisor ; then multiply the divisor, as it now stands, by the part of the root last obtained, for the subtra- hend. If there be more periods to be brought down, the operation must be repeated.

### Ex. 2.

Let the square root of 611524 be required.

$$611524(782$$
$$49$$
$$\overline{\phantom{000}}$$
$$148)1215$$
$$1184$$
$$\overline{\phantom{0000}}$$
$$1562)\ \ 3124$$
$$3124$$
$$\overline{\phantom{0000}}$$
$$*$$
$$\overline{\phantom{0000}}$$

(131.) In

(131.) In extracting the square root of a decimal, the pointing must be made the contrary way, beginning with the place of hundredths, or care must be taken to have an even number of decimal places; because, if the root have 1, 2, 3, 4, &c. decimal places, the square must have 2, 4, 6, 8, &c. places (Art. 38).

### Ex. 3.

To extract the square root of 64.853.

$$64.85\dot{3}\dot{0}(8.053 \text{ \&c.}$$
$$64$$

$$1605) \quad 8530$$
$$\phantom{1605)} \quad 8025$$

$$16103) \quad 50500$$
$$\phantom{16103)} \quad 48309$$

$$\phantom{16103)} \quad 2191 \text{ \&c.}$$

For every pair of cyphers which we suppose annexed to the decimal, another figure is obtained in the root.

(132.) The cube root of 1000 is 10, of 1000000 is 100, &c. therefore the cube root of a number less than 1000 consists of one figure, of any number between 1000 and 1000000, of two places of figures, &c. If then a point be made over every third figure contained in any number, beginning with the units, the number of points will shew the number of places in it's cube root.

Let the cube root of 405224 be required.

$$405224$$

$$405\overset{\cdot}{2}2\overset{\cdot}{4}(70 + 4$$
$$a^3 = 343000$$

$3a^2 = 14700) \; \overline{62224}$ the first remainder.

$$58800 = 3\,a^2b$$
$$3360 = 3\,ab^2$$
$$64 = b^3$$

$\overline{62224}$ subtrahend.

By pointing the number according to the direction, it appears that the root consists of two places ; let $a$ be the value of the figure in the ten's place, and $b$, of that in the unit's. Then $a$ is the greatest number whose cube is contained in 405000*, or 70; subtract it's cube from the whole quantity, and the remainder is 62224 ; divide this remainder by $3\,a^2$, or 14700, and the quotient 4, or $b$, is the second term in the root: then subtract the cube of 74 from the original number, and as the remainder is nothing, 74 is the cube root required. Observe, that the cyphers may be omitted in the operation ; and that as $a^3$ was at first subtracted, if from the first remainder, $3\,a^2b + 3\,ab^2 + b^3$ be taken, the whole cube of $a + b$ will be taken from the original quantity.

(133.) In extracting the cube root of a decimal, care must be taken that the decimal places be three, or some multiple of three, before the operation is begun ; because there are three times as many decimal places in the cube as there are in the root (Art. 38).

### Ex. 2.

Required the cube root of 311897.91.

311897.910

---

* See Art. 129.

$$3\overset{..}{1}189\overset{.}{7}.9\overset{.}{1}0(67.8$$
$$216... = a^3$$

$3a^3 = 108$ . . ) $95897$ first remainder

$$756 .. = 3\,a^2b$$
$$882 . = 3\,ab^2$$
$$343 = b^3$$

$84763$ subtrahend

$3a^2 = 13467$ . . )$11134910$ second remainder.

The new value of $a$ is $670$, or, omitting the cypher, $67$, and $3a^2$, the new divisor, is $13467$ . . hence 8 is the next figure in the root; and

$$107736 .. = 3\,a^2b$$
$$12864 . = 3\,ab^2$$
$$512 = b^3$$

$10902752$ subtrahend

$232158$ the third remainder.

It appears from the pointing, that there is one decimal place in the root; therefore $67.8$ is the root required, nearly. If three more cyphers be annexed to the decimal, another decimal place is obtained in the root; and thus approximation may be made to the true root of the proposed number, to any degree of accuracy.

Since the first remainder is $3\,a^2b + 3\,ab^2 + b^3$, the exact value of $b$ is not obtained by dividing by $3\,a^2$, and if upon trial the subtrahend be found to be greater than the first remainder, the value assumed for $b$ is too great, and a less number must be tried. The

greater

greater $a$ is with respect to $b$, the more nearly is the true value obtained by division; and when a few places in the root are found, the number of figures may nearly be doubled, by division only.

## ON SIMPLE EQUATIONS.

(134.) If one quantity be equal to another, or to nothing, and this equality be expressed algebraically, it constitutes an *Equation*.

Thus, $x-a=b-x$ is an equation, of which $x-a$ forms one side, and $b-x$ the other.

(135.) When an equation is cleared of fractions and surds, if it contain the first power only of an unknown quantity, it is called a *simple equation*, or an equation of one dimension: if the *square* of the unknown quantity be in any term, it is called a *quadractic*, or an equation of two dimensions; and in general, if the index of the highest power of the unknown quantity be $n$, it is called *an equation of* n *dimensions*.

(136.) *In any equation, quantities may be transposed from one side to the other, if their signs be changed, and the two sides will still be equal.*

Let $x+10=15$, then by subtracting 10 from each side, $x+10-10=15-10$ (Art. 68), or $x=15-10$.

Let $x-4=6$, by adding 4 to each side, $x-4+4=6+4$, or $x=6+4$ (Art. 67).

If $x-a+b=y$; adding $a-b$ to each side, $x-a+b+a-b=y+a-b$; or $x=y+a-b$.

(137.) Cor. Hence, if the signs of *all* the terms on each side be changed, the two sides will still be equal.

Let

Let $x-a=b-2x$; by transposition, $-b+2x=-x+a$; or $a-x=2x-b$.

(138.) *If every term, on each side, be multiplied by the same quantity, the results will be equal* (Art. 69).

(139.) Cor. An equation may be cleared of fractions, by multiplying every term, successively, by the denominators of those fractions.

Let $3x+\dfrac{5x}{4}=34$; multiplying by 4, $12x+5x=$ 136. (See Art. 103).

An equation may be cleared of fractions at once, by multiplying both sides by the product of all the denominators, or by any quantity which is a multiple of them all*.

Let $\dfrac{x}{2}+\dfrac{x}{3}+\dfrac{x}{4}=13$; multiplying by $2\times3\times4$, $3\times4\times x+2\times4\times x+2\times3\times x=2\times3\times4\times13$, or $12x+8x+6x=312$; that is, $26x=312$.

If each side be multiplied by 12, which is a multiple of 2, 3, and 4, the equation will become $\dfrac{12x}{2}+$ $\dfrac{12x}{3}+\dfrac{12x}{4}=156$; or $6x+4x+3x=156$; that is, $13x=156$.

(140.) *If each side of an equation be divided by the same quantity, the results will be equal.*

Let $17x=136$; then $x=\dfrac{136}{17}=8$ (Art. 70).

(141.) *If*

---

* If the *least* common multiple be made use of, the equation will be in the lowest terms.

(141.) *If each side of an equation be raised to the same power, the results will be equal.*

Let $x^{\frac{1}{2}}=9$ ; then $x=9 \times 9=81$ (Art. 69).

Also, if the same root be extracted on both sides, the results will be equal.

Let $x=81$ ; then $x^{\frac{1}{2}}=9$ (Art. 118).

(142.) *To find the value of an unknown quantity in a simple equation.*

Let the equation first be cleared of fractions, then transpose all the terms which involve the unknown quantity to one side of the equation, and the known quantities to the other; divide both sides by the co-efficient, or sum of the coefficients, of the unknown quantity, and the value required is obtained.

### Ex. 1.

To find the value of $x$ in the equation $3x - 5 = 23 - x$.

by transp. $3x + x = 23 + 5$  (Art. 136)

or $4x = 28$

by division $x = \dfrac{28}{4} = 7$ (Art. 140).

### Ex. 2.

Let $x + \dfrac{x}{2} - \dfrac{x}{3} = 4x - 17$.

Mult. by 2, and $2x + x - \dfrac{2x}{3} = 8x - 34$

Mult. by 3, and $6x + 3x - 2x = 24x - 102$ (Art. 139)

by transp.    $6x + 3x - 2x - 24x = -102$

or        $-17x = -102$

$17x = 102$ (Art. 137)

$$x = \frac{102}{17} = 6.$$

Ex. 3.

## Ex. 3.

$$\frac{1}{a} + \frac{b}{x} = c.$$

$$1 + \frac{ba}{x} = ca$$

$$x + ba = cax$$

$$x - cax = -ba$$

or $cax - x = ba$ (Art. 137)

i. e. $\overline{ca - 1}.x = ba$

$$x = \frac{ba}{ca - 1}.$$

## Ex. 4.

$$5 - \frac{x + 4}{11} = x - 3.$$

$$55 - x - 4 = 11x - 33$$

$$55 - 4 + 33 = 11x + x$$

$$84 = 12x$$

$$x = \frac{84}{12} = 7.$$

## Ex. 5.

$$x + \frac{3x - 5}{2} = 12 - \frac{2x - 4}{3}$$

$$2x + 3x - 5 = 24 - \frac{4x - 8}{3}$$

$$6x + 9x - 15 = 72 - 4x + 8$$

$$6x + 9x + 4x = 72 + 8 + 15$$

$$19x = 95$$

$$x = \frac{95}{19} = 5.$$

(143.) If

(143.) If there be two independent simple equations involving two unknown quantities, they may be reduced to one; which involves only one of the unknown quantities, by any of the following methods :

1ˢᵗ Method. In either equation, find the value of one of the unknown quantities in terms of the other and known quantities, and for it substitute this value in the other equation, which will then only contain one unknown quantity, whose value may be found by the rules before laid down.

$$\text{Let } \begin{cases} x + y = 10 \\ 2x - 3y = 5 \end{cases} \text{ To find } x \text{ and } y.$$

From the first equat. $x = 10 - y$; hence, $2x = 20 - 2y$,

$$\text{by subst. } 20 - 2y - 3y = 5$$
$$20 - 5 = 2y + 3y$$
$$15 = 5y$$
$$y = \frac{15}{5} = 3$$

hence also, $x = 10 - y = 10 - 3 = 7$.

2ᵈ Method. Find an expression for one of the unknown quantities, in each equation; put these expressions equal to each other, and from the resulting equation the other unknown quantity may be found.

$$\text{Let. } \begin{cases} x + y = a \\ bx + cy = de \end{cases} \text{ To find } x \text{ and } y.$$

From the first equat. $x = a - y$

from the second, $bx = de - cy$, and $x = \dfrac{de - cy}{b}$

therefore, $a - y = \dfrac{de - cy}{b}$

$$ba - by$$

$$ba - by = de - cy$$
$$\underline{cy - by = de - ba}$$
$$\overline{c - b} \cdot y = de - ba$$
$$y = \frac{de - ba}{c - b} \cdot$$

Also, $x = a - y$; that is,

$$x = a - \frac{de - ba}{c - b} = \frac{ca - ba - de + ba}{c - b} = \frac{ca - de}{c - b} \cdot$$

3$^d$ Method. If either of the unknown quantities have the same coefficient in both equations, it may be exterminated by subtracting, or adding, the equations, according as the sign of the unknown quantity, in the two cases, is the same or different.

Let $\begin{Bmatrix} x + y = 15 \\ x - y = 7 \end{Bmatrix}$ To find $x$ and $y$.

By subtraction, $2y = 8$, and $y = 4$

By addition, $2x = 22$, and $x = 11$ (Art. 67).

If the coefficients of the unknown quantity to be exterminated be different, multiply the terms of the first equation by the coefficient of the unknown quantity in the second, and the terms of the second equation by the coefficient of the same unknown quantity, in the first; then add, or subtract, the resulting equations, as in the former case.

Ex. 1. Let $\begin{Bmatrix} 3x - 5y = 13 \\ 2x + 7y = 81 \end{Bmatrix}$ To find $x$ and $y$.

Multiply the terms of the first equation by 2, and the terms of the other by 3,

$$\text{then } 6x - 10y = 26$$
$$6x + 21y = 243$$

By

By subtraction, $- 31y = - 217$

and $y = \dfrac{217}{31} = 7$ ;

also, $3x - 5y = 13$, or $3x - 35 = 13$

therefore, $3x = 13 + 35 = 48$

and $x = \dfrac{48}{3} = 16.$

### Ex. 2.

Let $\begin{cases} ax + by = c \\ mx - ny = d \end{cases}$ To find $x$ and $y$.

From the first, $max + mby = mc$

from the other, $max - nay = ad$

by subtraction, $mby + nay = mc - ad$

therefore, $y = \dfrac{mc - ad}{mb + na}.$

Again, $nax + nby = nc$

$mbx - nby = bd$

by addition, $\overline{na + mb}.x = nc + bd$

therefore, $x = \dfrac{nc + bd}{na + mb}.$

### Ex. 3.

Let $\begin{cases} \dfrac{3x - 5y}{2} + 3 = \dfrac{2x + y}{5} \\ 8 - \dfrac{x - 2y}{4} = \dfrac{x}{2} + \dfrac{y}{3} \end{cases}$ To find $x$ and $y$.

From the first equat. $3x - 5y + 6 = \dfrac{4x + 2y}{5}$

$15x - 25y + 30 = 4x + 2y$

$15x - 4x - 25y - 2y = - 30$

$11x - 27y = - 30$

from

from the second equat. $32 - x + 2y = \dfrac{4x}{2} + \dfrac{4y}{3} = 2x + \dfrac{4y}{3}$

$$96 - 3x + 6y = 6x + 4y$$
$$96 = 6x + 3x + 4y - 6y$$

or $9x - 2y = 96$

and $11x - 27y = -30$

hence $99x - 22y = 1056$

and $99x - 243y = -270$

$$221y = 1056 + 270 = 1326$$

$$y = \frac{1326}{221} = 6$$

also, $9x - 2y = 96$

or $9x - 12 = 96$

$$9x = 96 + 12 = 108$$

$$x = \frac{108}{9} = 12.$$

(144.) If there be three independent simple equations, and three unknown quantities, reduce two of the equations to one, containing only two of the unknown quantities, by the preceding rules; then reduce the third equation and either of the former to one, containing the same two unknown quantities; and from the two equations thus obtained, the unknown quantities which they involve may be found. The third quantity may be found by substituting their values in any of the proposed equations.

### Ex.

Let $\begin{cases} 2x + 3y + 4z = 16 \\ 3x + 2y - 5z = 8 \\ 5x - 6y + 3z = 6 \end{cases}$ To find $x$, $y$ and $z$.

From

From the two first equat. $6x + 9y + 12z = 48$
$$6x + 4y - 10z = 16$$
by subtr. $5y + 22z = 32$
from the first and third, $10x + 15y + 20z = 80$
$$10x - 12y + 6z = 12$$
by subtr. $27y + 14z = 68$
and $5y + 22z = 32$
hence $135y + 70z = 340$
and $135y + 594z = 864$
by subtr. $524z = 524$
$$z = 1$$
$$5y + 22z = 32$$
that is, $5y + 22 = 32$
$$5y = 32 - 22 = 10$$
$$y = \frac{10}{5} = 2$$
$$2x + 3y + 4z = 16$$
that is, $2x + 6 + 4 = 16$
$$2x = 16 - 6 - 4 = 6$$
$$x = 3.$$

The same method may be applied to any number of simple equations.

(145.) That the unknown quantities may have definite values, there must be as many independent equations as unknown quantities. When there are *more* equations than unknown quantities, the value of any one of these quantities may be determined from different equations; and should the values, thus found, differ, the equations are incongruous; should they be the same, one or more of the equations are unnecessary. When there are *fewer* equations than unknown quantities, one of these quantities cannot be
found

found, but in terms which involve some of the rest, whose values may be assumed at pleasure; and in such cases the number of answers is indefinite.

Thus, if $x+y=a$, $x=a-y$; and assuming $y$ at pleasure, we obtain a value of $x$, such, that $x+y=a$.

These equations must also be independent, that is, not deducible one from another.

Let $x+y=a$, and $2x+2y=2a$; this latter equation being deducible from the former, it involves no different supposition, nor requires any thing more for it's truth, than that $x+y=a$ should be a just equation.

## PROBLEMS WHICH PRODUCE SIMPLE EQUATIONS.

(146.) From certain quantities which are known, to investigate others which have a given relation to them, is the business of Algebra.

When a question is proposed to be resolved, we must first consider fully it's meaning and conditions. Then substituting for such unknown quantities as appear most convenient, we must proceed as if they were already determined, and we wished to try whether they answer all the proposed conditions or not, till as many independent equations arise as we have assumed unknown quantities, which will always be the case if the question be properly limited (Art. 145); and by the solution of these equations, the quantities sought will be determined.

Prob. 1.

## PROB. 1.

A bankrupt owes $A$ twice as much as he owes $B$, and $C$ as much as he owes $A$ and $B$ together; out of £.300, which is to be divided amongst them, what must each receive?

Let $x$ represent what $B$ must receive;

then $2x =$ what $A$ must receive,

and $x + 2x$, or $3x$, = what $C$ must receive;

amongst them they receive £.300; therefore

$$x + 2x + 3x = 300$$

$$6x = 300$$

$$x = \frac{300}{6} = 50, \text{ what } B \text{ must receive}$$

$$2x = 100, \text{ what } A \text{ must receive}$$

$$3x = 150, \text{ what } C \text{ must receive.}$$

## PROB. 2.

To divide a line of 15 inches into two such parts, that one may be three fourths of the other.

Let $4x =$ one part,

then $3x =$ the other.

$$7x = 15, \text{ by the question,}$$

$$x = \frac{15}{7}$$

$$4x = \frac{60}{7} = 8\frac{4}{7}, \text{ one part,}$$

$$3x = \frac{45}{7} = 6\frac{3}{7}, \text{ the other.}$$

## PROB. 3.

If $A$ can perform a piece of work in 8 days, and $B$ in 10 days, in what time will they finish it together?

Let $x$ be the time required.

In

In one day, $A$ performs $\frac{1}{8}$ part of the work ; therefore in $x$ days, he performs $\frac{x}{8}$ parts of it; and in the same time, $B$ performs $\frac{x}{10}$ parts of it; and calling the work 1,

$$\frac{x}{8} + \frac{x}{10} = 1$$
$$10x + 8x = 80$$
$$18x = 80$$
$$x = \frac{80}{18} = 4\frac{8}{18} = 4\frac{4}{9} \text{ days.}$$

## PROB. 4.

A workman was employed for 60 days, on condition that for every day he worked he should receive 15 pence; and for every day he played he should forfeit 5 pence; at the end of the time he had 20 shillings to receive; required the number of days he worked.

Let $x$ be the number of days he worked,
then $60 - x$ is the number he played,

$15x$ his pay, in pence,
$300 - 5x$, sum forfeited,
$15x - 300 + 5x = 240$, by the question,
$20x = 240 + 300 = 540$
$x = 27$, the days he worked,
$60 - x = 33$, the days he played.

## PROB. 5.

How much rye, at four shillings and sixpence a bushel, must be mixed with 50 bushels of wheat, at

six

six shillings a bushel, that the mixture may be worth five shillings a bushel?

Let $x$ be the number of bushels required;

then $9x$ is the price of the rye in sixpences

$\quad\quad$ 600 the price of the wheat

$\quad\quad$ $\overline{50+x}$ . 10 the price of the mixture;

therefore, $9x+600=500+10x$

and $100=x$, the number of bushels required.

## Prob. 6.

$A$ and $B$ engage together in play; in the first game, $A$ wins as much as he had and four shillings more, and finds he has twice as much as $B$; in the second game, $B$ wins half as much as he had at first and one shilling more, and then it appears that he has three times as much as $A$; what sum had each at first?

$\quad\quad$ Let $x$ be what $A$ had, in shillings,

$\quad\quad\quad$ $y$ what $B$ had

$\quad\quad\quad\quad$ $2x+4$, what $A$ has after the 1$^{\text{st}}$ game

$\quad\quad\quad\quad$ $y-x-4$, what $B$ has

by the question, $2x+4=2y-2x-8$

$\quad\quad$ or $2y-4x=12$

$\quad\quad\quad$ $y-2x=6$

$\quad\quad$ also, $y-x-4+\dfrac{y}{2}+1$, what $B$ has after

the second game,

$$2x+4-\dfrac{y}{2}-1, \text{ what } A \text{ has;}$$

by the question, $y-x-4+\dfrac{y}{2}+1=6x+12-\dfrac{3y}{2}-3$

$\quad\quad$ or $2y-2x-8+y+2=12x+24-3y-6$

$\quad\quad$ hence $6y-14x=24$

$\quad\quad$ or $3y-7x=12$

$\quad\quad\quad\quad\quad\quad\quad\quad\quad\quad\quad\quad$ also,

also, $y - 2x = 6$
therefore, $3y - 6x = 18$
also, $3y - 7x = 12$
by subtraction, $x = 6$
$y - 2x = 6$, or $y - 12 = 6$
$y = 18$.

## PROB. 7.

A smuggler had a quantity of brandy which he expected would raise £9 : 18s. ; after he had sold 10 gallons, a revenue officer seized one third of the remainder, in consequence of which he makes only £8 : 2s.; required the number of gallons he had, and the price per gallon.

Let $x$ be the number of gallons ;

then $\dfrac{198}{x}$ is the price per gallon, in shillings,

$\dfrac{x - 10}{3}$ the quantity seized,

$\dfrac{x - 10}{3} \times \dfrac{198}{x}$ the value of the quantity seized,

which appears by the question to be 36 shillings ;

therefore, $\dfrac{x - 10}{3} \times \dfrac{198}{x} = 36$

$\overline{x - 10} \times 66 = 36x$
$66x - 660 = 36x$
$30x = 660$
$x = 22$, the number of gallons,
$\dfrac{198}{x} = \dfrac{198}{22} = 9$ shillings, the price per gallon.

## PROB. 8.

$A$ and $B$ play at bowls, and $A$ bets $B$ three shillings
to

to two upon every game ; after a certain number of games it appears, that $A$ has won three shillings ; but had he ventured to bet five shillings to two, and lost one game more out of the same number, he would have lost thirty shillings : how many games did they play ?

Let $x$ be the number of games $A$ won,

$y$ the number $B$ won,

then $2x$ is what $A$ won of $B$,

and $3y$ what $B$ won of $A$.

$2x - 3y = 3$, by the question ;

$\overline{x - 1} . 2, A$ would win on the 2$^d$ supposition,

$\overline{y + 1} . 5, B$ would win,

$5y + 5 - 2x + 2 = 30$, by the question,

or $5y - 2x = 30 - 5 - 2 = 23$

therefore, $5y - 2x = 23$

and $2x - 3y = 3$

by addition, $5y - 3y = 26$

$2y = 26$

$y = 13$

$2x = 3 + 3y = 3 + 39 = 42$

$x = 21$

$x + y = 34$, the number of games.

## PROB. 9.

A sum of money was divided equally amongst a certain number of persons ; had there been three more, each would have received one shilling less, and had they been two fewer, each would have received one shilling more than he did : required the number of persons, and what each received.

Let

Let $x$ be the number of persons,

$y$ the sum each received, in shillings;

then $xy$ is the sum divided,

and $\overline{x+3} \times \overline{y-1} = xy$

also $\overline{x-2} \times \overline{y+1} = xy$ } by the question;

therefore, $xy - x + 3y - 3 = xy$

or $-x + 3y = 3$

and $xy + x - 2y - 2 = xy$

or $x - 2y = 2$

also, $-x + 3y = 3$

therefore, $y = 5$

hence $x - 2y = x - 10 = 2$.

or $x = 12$.

## ON QUADRATIC EQUATIONS.

(147.) When the terms of an equation involve the square of the unknown quantity, but the first power does not appear, the value of the square is obtained by the preceding rules ; and by extracting the square root on both sides, the quantity itself is found.

### Ex. 1.

Let $5x^2 - 45 = 0$; to find $x$.

By trans. $5x^2 = 45$

$x^2 = 9$

therefore (Art. 141), $x = \sqrt{9} = \pm 3$.

The signs $+$ and $-$ are both prefixed to the root, because the square root of a quantity may be either positive or negative (Art. 122). The sign of $x$ may also be negative ; but still $x$ will be either equal to $+3$ or $-3$.

Ex. 2.

### Ex. 2.

Let $ax^2 = bcd$; to find $x$.

$$x^2 = \frac{bcd}{a}$$

$$x = \pm \sqrt{\frac{bcd}{a}}.$$

(148.) If both the first and second powers of the unknown quantity be found in an equation, arrange the terms according to the dimensions of the unknown quantity, beginning with the highest, and transpose the known quantities to the other side; then, if the square of the unknown quantity be affected with a coefficient, divide all the terms by this coefficient, and if it's sign be negative, change the signs of all the terms (Art. 137), that the equation may be reduced to this form, $x^2 \pm px = \pm q$. Then add to both sides the square of half the coefficient of the first power of the unknown quantity, by which means, the first side of the equation is made a complete square, and the other consists of known quantities; and by extracting the square root on both sides, a simple equation is obtained, from which the value of the unknown quantity may be found.

### Ex. 1.

Let $x^2 + px = q$; now, we know that $x^2 + px + \frac{p^2}{4}$ is the square $x + \frac{p}{2}$ (Art. 127); add therefore, $\frac{p^2}{4}$ to both sides, and we have

$$x^2 + px$$

$x^2 + px + \dfrac{p^2}{4} = q + \dfrac{p^2}{4}$; then by extracting the square root on both sides,

$x + \dfrac{p}{2} = \pm \sqrt{q + \dfrac{p^2}{4}}$, and by transposition

$x = -\dfrac{p}{2} \pm \sqrt{q + \dfrac{p^2}{4}}$.

In the same manner, if $x^2 - px = q$, $x$ is found to be

$\dfrac{p}{2} \pm \sqrt{q + \dfrac{p^2}{4}}$.

### Ex. 2.

Let $x^2 - 12x + 35 = 0$; to find $x$.

By transposition, $x^2 - 12x = -35$, and adding the square of 6 to both sides of the equation,

$$x^2 - 12x + 36 = 36 - 35 = 1;$$

then extracting the square root on both sides,

$$x - 6 = \pm 1$$

$$x = 6 \pm 1 = 7 \quad \text{or} \quad 5;$$ either of which, substituted for $x$ in the original equation, answers the condition, that is, makes the whole equal to nothing.

### Ex. 3.

Let $\dfrac{6}{x + 1} + \dfrac{2}{x} = 3$; to find $x$.

$6 + \dfrac{2x + 2}{x} = 3x + 3$

$6x + 2x + 2 = 3x^2 + 3x$

$3x^2 - 5x = 2$

$x^2 - \dfrac{5x}{3} = \dfrac{2}{3}$

$x^2 - \dfrac{5x}{3} + \dfrac{25}{36} = \dfrac{25}{36} + \dfrac{2}{3}$

$x$ —

$$x - \frac{5}{6} = \pm \sqrt{\frac{25}{36} + \frac{2}{3}} = \pm \sqrt{\frac{25 + 24}{36}} = \pm \sqrt{\frac{49}{36}}$$

$$x - \frac{5}{6} = \pm \frac{7}{6}$$

$$x = \frac{5 \pm 7}{6} = 2, \text{ or } - \frac{2}{6}.$$

In this example, $\frac{25}{36}$ and $\frac{2}{3}$ are to be reduced to a common denominator, and since 36 is a complete square, the most convenient method for the solution, is to multiply both the numerator and denominator of $\frac{2}{3}$ by 12, that the common denominator may be a square number (Art. 20).

### Ex. 4.

(149.) Let $x + \sqrt{5x + 10} = 8$; to find $x$.

By transposition, $\sqrt{5x + 10} = 8 - x$

squaring both sides, $5x + 10 = 64 - 16x + x^2$

$$x^2 - 21x = 10 - 64 = -54$$

completing the square, $x^2 - 21x + \dfrac{441}{4} = \dfrac{441}{4} - 54$

$$= \frac{441 - 216}{4}, \text{ or } x^2 - 21x + \frac{441}{4} = \frac{225}{4}$$

extracting the square root, $x - \dfrac{21}{2} = \pm \dfrac{15}{2}$

$$x = \frac{21 \pm 15}{2} = 3 \text{ or } 18.$$

By this process two values of $x$ are found; but on trial it appears, that 18 does not answer the condition of the equation, if we suppose that $\sqrt{5x + 10}$ represents the positive square root of

$$5x + 10.$$

$5x + 10$. The reason is, that $5x + 10$ is the square of $-\sqrt{5x+10}$ as well as of $+\sqrt{5x+10}$; thus by squaring both sides of the equation $\sqrt{5x+10} = 8 - x$, a new condition is introduced, and a new value of the unknown quantity corresponding to it, which had no place before. Here, 18 is the value which corresponds to the supposition that $x - \sqrt{5x+10} = 8$.

It should be particularly observed, that, since $+x \times +y$ is equal to $-x \times -y$, in the multiplication and involution of quantities, new values are always introduced, which, if not again excluded by the nature of the question, will appear in the final equation.

(150.) Every equation, where the unknown quantity is found in two terms, and it's index in one is twice as great as in the other, may be resolved in the same manner.

### Ex. 5.

Let $z + 4z^{\frac{1}{2}} = 21$.

$$z + 4z^{\frac{1}{2}} + 4 = 21 + 4 = 25$$
$$z^{\frac{1}{2}} + 2 = \pm 5$$
$$z^{\frac{1}{2}} = \pm 5 - 2 = 3, \text{ or } -7,$$

therefore $z = 9$, or 49. (See Art. 149).

### Ex. 6.

Let $x^{-1} + x^{-\frac{1}{2}} = 6$

$$x^{-1} + x^{-\frac{1}{2}} + \frac{1}{4} = 6 + \frac{1}{4} = \frac{25}{4}$$
$$x^{-\frac{1}{2}} + \frac{1}{2} = \pm \frac{5}{2}$$
$$x^{-\frac{1}{2}} = \frac{-1 \pm 5}{2} = 2, \text{ or } -3$$

and $x^{\frac{1}{2}} = \frac{1}{2}$, or $-\frac{1}{3}$.

F 2

Ex. 7.

### Ex. 7.

Let $y^4 - 6y^2 - 27 = 0$.

$$y^4 - 6y^2 = 27$$
$$y^4 - 6y^2 + 9 = 27 + 9 = 36$$
$$y^2 - 3 = \pm 6$$
$$y^2 = 3 \pm 6 = 9, \text{ or} - 3$$
$$y = \pm 3, \text{ or } \pm \sqrt{-3}.$$

### Ex. 8.

Let $y^6 + ry^3 + \dfrac{q^3}{27} = 0$.

$$y^6 + ry^3 = -\frac{q^3}{27}$$

$$y^6 + ry^3 + \frac{r^2}{4} = \frac{r^2}{4} - \frac{q^3}{27}$$

$$y^3 + \frac{r}{2} = \pm \sqrt{\frac{r^2}{4} - \frac{q^3}{27}}.$$

$$y^3 = -\frac{r}{2} \pm \sqrt{\frac{r^2}{4} - \frac{q^3}{27}}$$

$$y = \sqrt[3]{-\frac{r}{2} \pm \sqrt{\frac{r^2}{4} - \frac{q^3}{27}}}.$$

(151.) When there are more equations and unknown quantities than one, a single equation, involving only one of the unknown quantities, may sometimes be obtained by the rules laid down for the solution of simple equations ; and one of the unknown quantities being discovered, the others may be obtained by substituting it's value in the preceding equations.

### Ex. 9.

Let $\begin{cases} x - \dfrac{x - y}{2} = 4 \\ y - \dfrac{x + 3y}{x + 2} = 1 \end{cases}$ To find $x$ and $y$.

From

From the first equation, $2x - x + y = 8$

$$\text{or } x + y = 8$$

$$\text{and } x = 8 - y$$

from the 2$^\text{d}$ equation, $xy + 2y - x - 3y = x + 2$

$$\text{or } xy - 2x - y = 2$$

by substitution, $\overline{8 - y} \times y - 2 \times \overline{8 - y} - y = 2$

$$8y - y^2 - 16 + 2y - y = 2$$

$$9y - y^2 = 16 + 2 = 18$$

$$y^2 - 9y = -18$$

$$y^2 - 9y + \frac{81}{4} = \frac{81}{4} - 18 = \frac{9}{4}$$

$$y - \frac{9}{2} = \pm \frac{3}{2}$$

$$y = \frac{9 \pm 3}{2} = 6, \text{ or } 3$$

$$x = 8 - y = 2, \text{ or } 5.$$

The solution will often be rendered more simple by particular artifices, the proper application of which is best learned by experience.

### Ex. 10.

Let $\begin{cases} x^2 + y^2 = 65 \\ xy = 28 \end{cases}$ To find $x$ and $y$.

From the second equation, $2xy = 56$

and adding this to the first, $x^2 + 2xy + y^2 = 121$

subtracting it from the same, $x^2 - 2xy + y^2 = 9$

by extracting the sq. roots, $x + y = \pm 11$

$$\text{and } x - y = \pm 3$$

$$\text{therefore, } 2x = \pm 14$$

$$x = 7, \text{ or } -7$$

$$\text{and } y = 4, \text{ or } -4.$$

(152.) It may sometimes be of use to substitute for
one

one of the unknown quantities, the product of the other and a third unknown quantity *.

## Ex. 11.

Let $\begin{Bmatrix} x^2 + xy = 12 \\ xy - 2y^2 = 1 \end{Bmatrix}$ To find $x$ and $y$.

$$\text{Let } vy = x$$
$$\text{then } v^2y^2 + vy^2 = 12$$
$$\text{and } vy^2 - 2y^2 = 1$$

from the former, $y^2 = \dfrac{12}{v^2 + v}$,

from the latter, $y^2 = \dfrac{1}{v - 2}$

therefore, $\dfrac{12}{v^2 + v} = \dfrac{1}{v - 2}$

$$\text{or } v^2 + v = 12v - 24$$
$$v^2 - 11v = -24$$
$$v^2 - 11v + \dfrac{121}{4} = \dfrac{121}{4} - 24$$
$$= \dfrac{121 - 96}{4} = \dfrac{25}{4}$$

hence, $v - \dfrac{11}{2} = \pm \dfrac{5}{2}$

$$v = \dfrac{11 \pm 5}{2} = 8, \text{ or } 3$$

$$y^2 = \dfrac{1}{v - 2} = \dfrac{1}{6}, \text{ or } 1$$

$$y = \pm \dfrac{1}{\sqrt{6}}, \text{ or } \pm 1$$

$$x = vy = \pm \dfrac{8}{\sqrt{6}}, \text{ or } \pm 3.$$

---

* This substitution may be successfully applied whenever the sum of the dimensions of the unknown quantities, in every term of each equation, is the same.

(153.) The operation may sometimes be shortened by substituting for the unknown quantities, the sum and difference of two others *.

Ex. 12.

Let $\begin{cases} \dfrac{x^2}{y} + \dfrac{y^2}{x} = 18 \\ x + y = 12 \end{cases}$ To find $x$ and $y$.

Assume $x = z + v$

and $y = z - v$

then $x + y = 2z = 12$

or $z = 6$

hence, $x = 6 + v$

and $y = 6 - v$

also, since $\dfrac{x^2}{y} + \dfrac{y^2}{x} = 18$

$x^3 + y^3 = 18xy$

and $x^3 = \overline{6+v}|^3 = 216 + 108v + 18v^2 + v^3$

$y^3 = \overline{6-v}|^3 = 216 - 108v + 18v^2 - v^3$

therefore, $x^3 + y^3 = 432 + 36v^2$

$xy = \overline{6+v} \times \overline{6-v} = 36 - v^2$

$18xy = 648 - 18v^2$

but $x^3 + y^3 = 18xy$

therefore, $432 + 36v^2 = 648 - 18v^2$

$54v^2 = 216$

$v^2 = \dfrac{216}{54} = 4$

$v = \pm 2$

$x = 6 \pm 2 = 8$ or $4$

$y = 6 \mp 2 = 4$ or $8$ †

PROBLEMS

---

* This artifice may be used, when the unknown quantities in each equation are similarly involved.

† Other methods are given by Dr. Waring, *Med. Alg.* Cap. 4.

## PROBLEMS PRODUCING QUADRATIC EQUATIONS.

### PROB. 1.

(154.) A person bought a certain number of oxen for 80 guineas, and if he had bought 4 more for the same sum, they would have cost a guinea a piece less; required the number of oxen and price of each.

Let $x$ be the number,

then $\dfrac{80}{x}$ is the price of each,

and $\dfrac{80}{x+4}$ the price of each on the 2$^d$ supposition,

$\dfrac{80}{x+4} = \dfrac{80}{x} - 1$, by the question,

$80 = \dfrac{80x+320}{x} - x - 4$

$80x = 80x + 320 - x^2 - 4x$

$x^2 + 4x = 320$

$x^2 + 4x + 4 = 324$

$x + 2 = \pm 18$

$x = \pm 18 - 2 = 16$ or $- 20$.

$\dfrac{80}{x} = \dfrac{80}{16} = 5$ guineas, the price of each.

In this, and in many other cases, especially in the solution of philosophical questions, we deduce, from the algebraical process, answers which do not correspond with the conditions. The reason seems to be, that the algebraical expression is more general than the common language, and the equation which is a proper representation

representation of the conditions, will also express other conditions, and answer other suppositions. In the foregoing instance, $x$ may either represent a positive or a negative quantity, and cannot in the operation represent a positive quantity alone (Art. 149); and the equation $\dfrac{80}{x+4} = \dfrac{80}{x} - 1$, when $x$ is negative, or represents the diminution of stock, will be a proper expression for the solution of the following problem: A person sells a certain number of oxen for 80 guineas; and, had he sold 4 fewer for the same sum, he would have received a guinea a piece more for them; required the number sold.

## Prob. 2.

(155.) To divide a line of 20 inches into two such parts, that the rectangle under the whole and one part, may be equal to the square of the other.

Let $x$ be the greater part, then will $20 - x$ be the less, and $x^2 = \overline{20 - x} . 20 = 400 - 20x$, by the question,

$x^2 + 20x = 400$

$x^2 + 20x + 100 = 400 + 100 = 500$

$x + 10 = \pm \sqrt{500}$

$x = + \sqrt{500} - 10$, or $- \sqrt{500} - 10$.

The observation contained in the preceding article may be applied here; and it is to be remarked, that the negative values thus deduced are not insignificant, or useless. Here the negative value shews, that if the line be produced $\sqrt{500} + 10$ inches, the square of the part produced is equal to the rectangle under the line given, and the line made up of the whole and part produced.

Prob. 3.

## PROB. 3.

(156.) To find two numbers, whose sum, product, and the sum of whose squares, are equal to each other.

Let $x+y$ and $x-y$ be the numbers,

their sum is $2x$

their product $x^2 - y^2$

the sum of their squares $2x^2 + 2y^2$

and by the question $2x = 2x^2 + 2y^2$

$$\text{or } x = x^2 + y^2$$

$$\text{also, } 2x = x^2 - y^2$$

$$\text{therefore, } 3x = 2x^2$$

$$x = \frac{3}{2}$$

$$2x = x^2 - y^2$$

$$\text{or } 3 = \frac{9}{4} - y^2$$

$$y^2 = \frac{9}{4} - 3 = \frac{9-12}{4} = \frac{-3}{4}$$

$$y = \pm \frac{\sqrt{-3}}{2}$$

$$x+y = \frac{3 + \sqrt{-3}}{2}$$

$$x-y = \frac{3 - \sqrt{-3}}{2}.$$

Since the square of every quantity is positive, a negative quantity has no square root; the conclusion therefore shews that there are no such numbers as the question supposes *.

ON

---

* An excellent collection of Problems, producing simple and quadratic equations, has lately been published by the Rev. M. Bland.

## ON RATIOS.

(157.) Ratio is the relation which one quantity bears to another in respect of magnitude, the comparison being made by considering what multiple, part, or parts, one is of the other.

Thus, in comparing 6 with 3, we observe that it has a certain magnitude with respect to 3, which it contains twice; again, in comparing it with 2, we see that it has a different relative magnitude, for it contains 2 three times, or it is greater when compared with 2 than it is when compared with 3. The ratio of $a$ to $b$ is usually expressed by two points placed between them, thus, $a : b$; and the former is called the *antecedent* of the ratio, the latter the *consequent*.

(158.) Cor. 1. When one antecedent is the same multiple, part, or parts, of it's consequent, that another antecedent is of it's consequent, the ratios are equal. Thus, the ratio of 4 : 6 is equal to the ratio of 2 : 3, *i. e.* 4 has the same magnitude when compared with 6, that 2 has when compared with 3, since $\frac{4}{6} = \frac{2}{3}$; the ratio of $a : b$ is equal to the ratio of $c : d$, if $\frac{a}{b} = \frac{c}{d}$, because $\frac{a}{b}$ and $\frac{c}{d}$, represent the multiple, part, or parts, that $a$ is of $b$, and $c$ of $d$.

(159.) Cor. 2. If the terms of a ratio be multiplied or divided by the same quantity, the ratio is not altered.

For $\frac{a}{b} = \frac{ma}{mb}$ (Art. 89).

(160.) Cor.

(160.) Cor. 3. That ratio is greater than another, whose antecedent is the greater multiple, part, or parts, of it's consequent. Thus, the ratio of 7 : 4 is greater than the ratio of 8 : 5; because $\frac{7}{4}$, or $\frac{35}{20}$ is greater than $\frac{8}{5}$, or $\frac{32}{20}$. These conclusions follow immediately from our idea of ratio.

(161.) A ratio is called a ratio *of greater inequality, of less inequality,* or *of equality,* according as the antecedent is *greater, less* than, or *equal* to, the consequent.

(162.) *A ratio of greater inequality is diminished, and of less inequality increased, by adding any quantity to both it's terms.*

If to the terms of the ratio 7 : 4, 1 be added, it becomes the ratio of 8 : 5, which is less than the former, (Art. 160). And in general, let $x$ be added to the terms of the ratio $a : b$, and it becomes $a + x : b + x$, which is greater or less than the former, according as $\frac{a+x}{b+x}$ is greater or less than $\frac{a}{b}$ ; or, by reducing them to a common denominator, as $\frac{ab+bx}{b.\overline{b+x}}$ is greater or less than $\frac{ab+ax}{b.\overline{b+x}}$ ; that is, as $b$ is greater or less than $a$.

(163.) Hence, a ratio of greater inequality is increased, and of less inequality diminished, by taking from the terms a quantity less than either of them.

(164.) If the antecedents of any ratios be multiplied together, and also the consequents, a new ratio results, which is *said* to be compounded of the former.
Thus,

Thus, $ac : bd$ is said to be compounded of the two
$a : b$ and $c : d$.  It is also sometimes *called* the sum
of the ratios; and when the ratio $a : b$ is compounded
with itself, the resulting ratio, $a^2 : b^2$, is *called* the
double of the ratio of $a : b$, and if three of these ratios
be compounded together, the result $a^3 : b^3$, is *called*
the triple of the first, &c.   Also, the ratio of $a : b$ is
*said* to be one third of the ratio of $a^3 : b^3$; and
$a^{\frac{1}{m}} : b^{\frac{1}{m}}$ is said to be an $m^{th}$ part of the ratio of $a : b$.

(165.) Let the first ratio be $a : 1$; then $a^2 : 1, a^3 : 1$,
....$a^n : 1$, are twice, three times, ...$n$ times the first
ratio; where $n$, the index of $a$, shews what multiple,
or part, of the ratio $a^n : 1$, the first ratio $a : 1$, is.
On this account, the indices 1, 2, 3, ...$n$, are called
measures of the ratios $a^1 : 1, a^2 : 1, a^3 : 1, \ldots .$
$a^n : 1$.

(166.) *If the consequent of the preceding ratio be
the antecedent of the succeeding one, and any number
of such ratios be taken, the ratio which arises from
their composition, is that of the first antecedent to
the last consequent.*

Let $a : b, b : c, c : d$, &c. be the ratios, the com-
pound ratio is $a \times b \times c : b \times c \times d$, (Art. 164.); or,
dividing by $b \times c$ (Art. 159), $a : d$.

(167.) *A ratio of greater inequality, compounded
with another, increases it; and a ratio of less inequality diminishes it.*

Let the ratio of $x : y$ be compounded with the ratio
of $a : b$, and the resulting ratio $ax : by$ is greater or
less than the ratio $a : b$, according as $\dfrac{ax}{by}$ is greater or

less

less than $\dfrac{a}{b}$ (Art. 160) ; i. e. according as $x$ is greater or less than $y$.

(168.) *If the difference between the antecedent and consequent of a ratio be small when compared with either of them, the double of the ratio, or the ratio of their squares, is nearly obtained by doubling this difference.*

Let $a + x : a$ be the proposed ratio, where $x$ is small when compared with $a$; then $a^2 + 2ax + x^2 : a^2$ is the ratio of the squares of the antecedent and consequent; but since $x$ is small when compared with $a$, $x^2$ or $x \times x$ is small when compared with $2a \times x$, and much smaller than $a \times a$; therefore, $a^2 + 2ax : a^2$, or $a + 2x : a$ (Art. 159), will nearly express the ratio of $a^2 + 2ax + x^2 : a^2$.

Thus, the ratio of the square of 1001 to the square of 1000 is nearly 1002 : 1000 ; the real ratio is 1002.001 : 1000, in which the antecedent differs from it's approximate value, only by the thousandth part of an unit.

(169.) Cor. Hence, the ratio of the square root of $a + 2x$ to the square root of $a$ is the ratio $a + x : a$, nearly; that is, if the difference of two quantities be small with respect to either of them, the ratio of their square roots is nearly obtained by halving their difference.

(170.) In the same manner, $a + 3x : a$; $a + 4x : a$; $a + mx : a$; are nearly equal to the ratios $\overline{a + x}|^3 : a^3$; $\overline{a + x}|^4 : a^4$; $\overline{a + x}|^m : a^m$; if $mx$ be small when compared with $a$.

ON

## ON PROPORTION.

(171.) Four quantities are said to be *proportionals*, when the first is the same multiple, part, or parts, of the second, that the third is of the fourth. That is, when $\frac{a}{b} = \frac{c}{d}$, the four quantities $a$, $b$, $c$, $d$, are called proportionals. This is usually expressed by saying, $a$ is to $b$ as $c$ to $d$; and thus represented, $a : b :: c : d$.

The terms $a$ and $d$ are called the *extremes*, and $b$ and $c$ the *means*.

(172.) *When four quantities are proportionals, the product of the extremes is equal to the product of the means.*

Let $a$, $b$, $c$, $d$, be the four quantities; then, since they are proportionals, $\frac{a}{b} = \frac{c}{d}$ (Art. 171); and by multiplying both sides by $bd$, $ad = bc$.

(173.) Cor. 1. If the first be to the second as the second to the third, the product of the extremes is equal to the square of the mean.

(174.) Cor. 2. Any three terms in a proportion being given, the fourth may be determined from the equation $ad = bc$.

(175.) *If the product of two quantities be equal to the product of two others, the four are proportionals, making the terms of one product the means, and the terms of the other, the extremes.*

Let $xy = ab$, then dividing by $ay$, $\frac{x}{a} = \frac{b}{y}$, or $x : a :: b : y$. (Art. 171).

(176.) *If*

(176.) *If* a : b :: c : d, *and* c : d :: e : f, *then will* a : b :: e : f.

Because $\frac{a}{b} = \frac{c}{d}$, and $\frac{c}{d} = \frac{e}{f}$; therefore, $\frac{a}{b} = \frac{e}{f}$; or $a :$ $b :: e : f$.

(177.) *If four quantities be proportionals, they are also proportionals when taken* inversely.

If $a : b :: c : d$, then $b : a :: d : c$. For $\frac{a}{b} = \frac{c}{d}$, and dividing unity by each of these equal quantities, or taking their reciprocals, $\frac{b}{a} = \frac{d}{c}$; that is, $b :$ $a :: d : c$.

(178.) *If four quantities be proportionals, they are proportionals when taken* alternately.

If $a : b :: c : d$, then $a : c :: b : d$.

Because the quantities are proportionals, $\frac{a}{b} = \frac{c}{d}$; and multiplying by $\frac{b}{c}$, $\frac{a}{c} = \frac{b}{d}$, or $a : c :: b : d$.

Unless the four quantities are of the *same* kind, the alternation cannot take place, because this operation supposes the first to be some multiple, part, or parts, of the third.

One line may have to another line the same ratio that one weight has to another weight, but a line has no relation, in respect of magnitude to a weight. In cases of this kind, if the four quantities be represented by numbers, or other quantities which are similar, the alternation may take place, and the conclusions drawn from it will be just.

(179.) *When*

(179.) *When four quantities are proportionals, the first together with the second, is to the second, as the third together with the fourth, is to the fourth.*

Let $a : b :: c : d$, then componendo, $a + b : b :: c + d : d$.

Because $\frac{a}{b} = \frac{c}{d}$, by adding unity to each side, $\frac{a}{b} + 1$

$= \frac{c}{d} + 1$ ; that is, $\frac{a+b}{b} = \frac{c+d}{d}$ ; or, $a + b : b :: c + d : d$.

(180.) *Also,* dividendo, *the excess of the first above the second, is to the second, as the excess of the third above the fourth, is to the fourth.*

Because $\frac{a}{b} = \frac{c}{d}$, by subtracting unity from each

side, $\frac{a}{b} - 1 = \frac{c}{d} - 1$ ; that is, $\frac{a-b}{b} = \frac{c-d}{d}$; or, $a - b$ $: b :: c - d : d$.

(181.) *Again,* convertendo, *the first is to it's excess above the second, as the third to it's excess above the fourth.*

By the last article, $\frac{a-b}{b} = \frac{c-d}{d}$; and $\frac{b}{a} = \frac{d}{c}$ (Art.

177) ; therefore, $\frac{a-b}{b} \times \frac{b}{a} = \frac{c-d}{d} \times \frac{d}{c}$ ; or, $\frac{a-b}{a} =$

$\frac{c-d}{c}$, that is, $a - b : a :: c - d : c$; and inversely,

$a : a - b :: c : c - d$.

(182.) *When four quantities are proportionals, the sum of the first and second is to their difference, as the sum of the third and fourth, to their difference.*

Let $a : b :: c : d$; then, $a + b : a - b :: c + d : c - d$.

By

By Art. 179, $\dfrac{a+b}{b} = \dfrac{c+d}{d}$; and by Art. 180, $\dfrac{a-b}{b}$

$= \dfrac{c-d}{d}$; therefore, $\dfrac{a+b}{b} \div \dfrac{a-b}{b} = \dfrac{c+d}{d} \div \dfrac{c-d}{d}$

(Art. 70); or, $\dfrac{a+b}{a-b} = \dfrac{c+d}{c-d}$; that is, $a+b : a-b ::$ $c+d : c-d$.

(183.) *When any number of quantities are proportionals, as one antecedent is to it's consequent, so is the sum of all the antecedents, to the sum of all the consequents.*

Let $a : b :: c : d :: e : f$, &c.
then $a : b :: a+c+e : b+d+f$.

Because $\dfrac{a}{b} = \dfrac{c}{d}$, $ad = bc$; in the same manner, $af = be$; also, $ab = ba$; hence, $ab + ad + af = ba + bc + be$, or, $a.\overline{b+d+f} = b.\overline{a+c+e}$; and by Art. 175, $a : b ::$ $a+c+e : b+d+f$.

(184.) *When four quantities are proportionals, if the first and second be multiplied, or divided, by any quantity, as also the third and fourth, the resulting quantities will be proportionals.*

Let $a : b :: c : d$, then will $ma : mb :: \dfrac{c}{n} : \dfrac{d}{n}$.

For $\dfrac{a}{b} = \dfrac{c}{d}$; therefore, $\dfrac{ma}{mb} = \dfrac{\frac{1}{n} . c}{\frac{1}{n} . d}$ (Art. 89);

or, $ma : mb :: \dfrac{c}{n} : \dfrac{d}{n}$.

(185.) *If the first and third be multiplied, or divided, by any quantity, and also the second and fourth, the resulting quantities will be proportionals.*

For

For $\dfrac{a}{b} = \dfrac{c}{d}$; therefore $\dfrac{ma}{b} = \dfrac{mc}{d}$; and $\dfrac{ma}{\frac{1}{n}.b} = \dfrac{mc}{\frac{1}{n}.d}$

(Art. 69); or, $ma : \dfrac{b}{n} :: mc : \dfrac{d}{n}$.

(186.) Cor. Hence, in any proportion, if instead of the second and fourth terms, quantities proportional to them be substituted, we have still a proportion. For $\dfrac{b}{n}$ and $\dfrac{d}{n}$ are in the same proportion with $b$ and $d$ (Art. 184.)

(187.) *In two ranks of proportionals, if the corresponding terms be multiplied together, the products will be proportionals.*

Let $a : b :: c : d$
and $e : f :: g : h$
then will $ae : bf :: cg : dh$

Because $\dfrac{a}{b}=\dfrac{c}{d}$, and $\dfrac{e}{f}=\dfrac{g}{h}$; therefore, $\dfrac{ae}{bf}=\dfrac{cg}{dh}$; or,
$ae : bf :: cg : dh$.
This is called *compounding* the proportions.

The proposition is true if applied to any number of proportions.

(188.) *If four quantities be proportionals, the like powers, or roots of these quantities, will be proportionals.*

Let $a : b :: c : d$, then $\dfrac{a}{b} = \dfrac{c}{d}$, and $\dfrac{a^n}{b^n} = \dfrac{c^n}{d^n}$; or,
$a^n : b^n :: c^n : d^n$; where $n$ is whole or fractional.

+ (189.) *If two numbers, a and* b, *be prime to each other, they are the least in that proportion.*

If

If possible, let $\frac{a}{b} = \frac{c}{d}$, where $a$ and $b$ are prime to each other, and respectively greater than $c$ and $d$. If the latter numbers be not prime to each other, divide them by their greatest common measure. Then divide $a$ by $b$, and $c$ by $d$, as in Art. 90; thus,

$$b) \, a \, (m \qquad\qquad\qquad d) \, c \, (m$$
$$\overline{\qquad\qquad}$$
$$x) \, b \, (n \qquad\qquad\qquad r) \, d \, (n$$
$$\overline{\qquad\qquad}$$
$$y \qquad\qquad\qquad\qquad s$$

and because $\frac{a}{b} = \frac{c}{d}$, the first quotients $m$, $m$, are equal; again, since $\frac{a}{b} = m + \frac{x}{b}$, and $\frac{c}{d} = m + \frac{r}{d}$, we have $\frac{x}{b} = \frac{r}{d}$, or $\frac{b}{x} = \frac{d}{r}$; also, because $b$ is greater than $d$, $x$ is greater than $r$. In the same manner, $\frac{x}{y} = \frac{r}{s}$, and $y$ is greater than $s$, &c. thus the remainder in the latter division will become unity, sooner than the remainder in the former. Let $s = 1$; then $\frac{x}{y} = r$, and $y$, which is greater than unity, will measure $a$ and $b$ (Art. 92), which is contrary to the supposition.

COR. Hence, if $\frac{a}{b} = \frac{c}{d}$, and $a$ and $b$ be prime to each other, $c$ and $d$ are equimultiples of $a$ and $b$.

(190.) If $a$ and $b$ be each of them prime to $c$, $ab$ is prime to $c$.

If not, let $ab = mr$, and $c = ms$; then since $a$ and $b$ are prime to $c$, they are respectively *prime* to $ms$, and therefore to $m$; and because $ab = mr$, we have $\frac{a}{m} = \frac{r}{b}$;

therefore

therefore $b$ is a *multiple* of $m$ (Art. 189. Cor.), which is absurd.

Cor. 1. If $b$ be equal to $a$, then $a^2$ and $c$ have no common measure; or $\dfrac{a^2}{c}$ is a fraction in it's lowest terms.

Cor. 2. In the same manner, $\dfrac{a^3}{c}$, $\dfrac{a^4}{c}$, &c. are fractions in their lowest terms.

Cor. 3. If $a$, $b$, and $c$, be *each* of them prime to $d$, $e$, and $f$, $abc$ is prime to $def$.

For, if $a$ be prime to $d$ and $e$, it is prime to $de$, and if it be prime to $de$ and $f$, it is prime to $def$. In the same manner, $b$ and $c$ are prime to $def$; consequently, $abc$ is prime to $def$.

Cor. 4. If $a$ be prime to $b$, $a^2$ is prime to $b^2$, and $a^3$ to $b^3$, &c.

## SCHOLIUM.

(191.) In the definition of Proportion, it is supposed that one quantity is some determinate multiple, part, or parts of another; or that the fraction arising from the division of one by the other, (which expresses the multiple, part, or parts, that the former is of the latter), is a determinate fraction. This will be the case, whenever the two quantities have any common measure whatever.

Let $x$ be a common measure of $a$ and $b$, and let $a = mx$, $b = nx$; then $\dfrac{a}{b} = \dfrac{mx}{nx} = \dfrac{m}{n}$, where $m$ and $n$ are whole numbers.

But it sometimes happens that the quantities are
*incom-*

*incommensurable*, or admit of no common measure whatever, as when one represents the circumference of a circle and the other it's diameter ; in such cases, the value of $\frac{a}{b}$ cannot be exactly expressed by any fraction, $\frac{m}{n}$, whose numerator and denominator are whole numbers; yet a fraction of this kind may be found, which will express it's value to any required degree of accuracy.

Suppose $x$ to be a measure of $b$, and let $b = nx$ ; also let $a$ be greater than $mx$ but less than $\overline{m+1}.x$ ; then $\frac{a}{b}$ is greater than $\frac{m}{n}$ but less than $\frac{m+1}{n}$, or the difference between $\frac{m}{n}$ and $\frac{a}{b}$ is less than $\frac{1}{n}$ ; and as $x$ is diminished, since $nx = b$, $n$ is increased, and $\frac{1}{n}$ diminished ; therefore by diminishing $x$, the difference between $\frac{m}{n}$ and $\frac{a}{b}$ may be made less than any that can be assigned.

If $a$ and $b$ as well as $c$ and $d$ be incommensurable, and if when $\frac{a}{b}$ lies between $\frac{m}{n}$ and $\frac{m+1}{n}$, $\frac{c}{d}$ lie also between $\frac{m}{n}$ and $\frac{m+1}{n}$, however the magnitudes $m$ and $n$ are increased, $\frac{a}{b}$ is equal to $\frac{c}{d}$. If they are not equal, they must have some assignable difference, and because each of them lies between $\frac{m}{n}$ and $\frac{m+1}{n}$, this difference

is

is less than $\frac{1}{n}$; but since $n$ may, by the supposition, be increased without limit, $\frac{1}{n}$ may be diminished without limit, that is, it may become less than any assignable magnitude; therefore, $\frac{a}{b}$ and $\frac{c}{d}$ have no assignable difference; that is, $\frac{a}{b}$ is equal to $\frac{c}{d}$; and all the preceding propositions, respecting proportionals, are true of the four magnitudes, $a, b, c, d.$

## ON VARIABLE QUANTITIES.

(192.) In the investigation of the relation which varying and dependent quantities bear to each other, the conclusions are more readily obtained, by expressing only two terms in each proportion, than by retaining the four.

But though, in considering the variation of such quantities, two terms only are expressed, it will be necessary for the Learner to keep constantly in mind that four are supposed; and that the operations, by which our conclusions are in this case obtained, are in reality the operations of proportionals.

(193.) DEF. 1. One quantity is said to *vary directly* as another, when the two quantities depend wholly upon each other, and in such a manner, that if one be changed, the other is changed in the same proportion.

Let $A$ and $B$ be mutually dependent upon each other, in such a way, that if $A$ be changed to any other value $a$, $B$ must be changed to another value $b$, such,

such, that $A : a :: B : b$, then $A$ is said to vary directly as $B$.

Ex. If the altitude of a triangle be invariable, the area varies as the base. For if the base be increased, or diminished, the area is increased or diminished in the same proportion*.

(194.) Def. 2. One quantity is said to *vary inversely* as another, when the former cannot be changed in any manner, but the reciprocal of the latter is changed in the same proportion.

$A$ varies inversely as $B$, $(A \propto \frac{1}{B})$, if, when $A$ is changed to $a$, $B$ be changed to $b$, in such a manner that $A : a :: \frac{1}{B} : \frac{1}{b}$; or $A : a :: b : B$.

Ex. If the area of a triangle be given, the base varies inversely as the perpendicular altitude.

Let $A$ and $a$ represent the altitudes, $B$ and $b$, the bases, of two equal triangles; then $\frac{A \times B}{2} = \frac{a \times b}{2}$ (the area of a triangle being the half of the rectangle under the base and perpendicular), or $A \times B = a \times b$; therefore (Art. 175), $A : a :: b : B :: \frac{1}{B} : \frac{1}{b}$.

(195.) Def. 3. One quantity is said to *vary as two others jointly*, if, when the former is changed in any manner, the product of the other two be changed in the same proportion.

Thus, $A$ varies as $B$ and $C$ jointly, $(A \propto BC)$, when $A$ cannot be changed to $a$, but the product $BC$ must be changed to $bc$, such, that $A : a :: BC : bc$.

Ex.

---

* The sign $\propto$ placed between two quantities signifies that they vary as each other.

Ex. The area of a triangle varies as it's base and perpendicular altitude jointly. Let $A$, $B$, $P$, represent the area, base, and perpendicular altitude of one triangle; $a$, $b$, $p$, those of another; then $BP = 2A$, and $bp = 2a$; therefore $\dfrac{A}{a} = \dfrac{BP}{bp}$, or $A : a :: BP : bp$.

(196.) Def. 4. One quantity is said to vary *directly* as a second and *inversely* as a third, when the first cannot be changed in any manner, but the second multiplied by the reciprocal of the third, is changed in the same proportion.

$A$ varies directly as $B$, and inversely as $C$, $\left( A \propto \dfrac{B}{C} \right)$, when $A : a :: \dfrac{B}{C} : \dfrac{b}{c}$; $A$, $B$, $C$, and $a$, $b$, $c$ being corresponding values of the three quantities.

Ex. The base of a triangle varies as the area directly and the perpendicular altitude inversely. The notation in the last Article being retained, $\dfrac{BP}{bp} = \dfrac{A}{a}$; and multiplying both sides by $\dfrac{p}{P}$, we have $\dfrac{B}{b} = \dfrac{Ap}{aP} = \dfrac{A}{P} \div \dfrac{a}{p}$; therefore, $B : b :: \dfrac{A}{P} : \dfrac{a}{p}$.

In the following articles, $A$, $B$, $C$, &c. represent corresponding values of any quantities, and, $a$, $b$, $c$, &c. any other corresponding values of the same quantities.

(197.) *If one quantity vary as a second, and that second as a third, the first varies as the third.*

Let $A : a :: B : b$, and $B : b :: C : c$, then, (Art. 176), $A : a :: C : c$. That is, $A \propto C$. In the same manner, if $A \propto B$ and $B \propto \dfrac{1}{C}$, then $A \propto \dfrac{1}{C}$.

(198.) *If*

(198.) *If two quantities vary respectively as a third, their sum, or difference, or the square root of their product, will vary as the third.*

Let $A \propto C$ and $B \propto C$, then $\overline{A \pm B} \propto C$; also, $\sqrt{AB}$ $\propto C$. By the supposition, $A : a :: C : c :: B : b$; there-fore $A : a :: B : b$; alternately, $A : B :: a : b$; and by composition or division, $A \pm B : B :: a \pm b : b$; alt. $A \pm B : a \pm b :: B : b :: C : c$; that is, $C \propto \overline{A \pm B}$.

Again, $A : a :: C : c$
and $B : b :: C : c$
therefore, $AB : ab :: C^2 : c^2$ (Art. 187)
and $\sqrt{AB} : \sqrt{ab} :: C : c$ (Art. 188); that is, $C \propto \sqrt{AB}$.

(199.) *If one quantity vary as another, it will also vary as any multiple, or part, of the other.*

Let $A \propto B$, and $m$ be any constant quantity, then because $A : a :: B : b$, $A : a :: mB : mb$, or $A : a :: \dfrac{B}{m}$ $: \dfrac{b}{m}$ (Art. 184); that is, $A \propto mB$ or $\propto \dfrac{B}{m}$.

(200.) Cor. 1. If $A$ vary as $B$, $A$ is equal to $B$ mul-tiplied by some invariable quantity. For $A : a :: mB$ $: mb$; altern. $A : mB :: a : mb$; if therefore $m$ be so assumed that $A = mB$, then in all cases, $a = mb$.

(201.) Cor. 2. If we know any corresponding values of $A$ and $B$, the constant quantity $m$ may be found.

Let $a$ and $b$ be the two values known, then $m = \dfrac{a}{b}$;

and in general, $A = \dfrac{a}{b} \times B$.

Ex.

Ex. Let $S \propto T^2$, and when $T = 1$ suppose $S = 16$, then $S = 16 T^2$.

(202.) *If one quantity vary as another, any power or root of the former will vary as the same power or root of the latter.*

Let $A$ vary as $B$, then $A : a :: B : b$, and by Art. 188, $A^i : a^n :: B^n : b^n$; that is, $A^n \propto B^n$, where $n$ is whole or fractional.

(203.) *If one quantity vary as another, and each of them be multiplied or divided by any quantity, variable or invariable, the products or quotients will vary as each other.*

Let $A$ vary as $B$, and let $T$ be any other quantity. Then, by the supposition, $A : a :: B : b$; therefore

$$AT : at :: BT : bt, \text{ and } \frac{A}{T} : \frac{a}{t} :: \frac{B}{T} : \frac{b}{t} \text{ (Art. 185).}$$

204.) Cor. If $A \propto B$, dividing both by $B$, $\dfrac{A}{B} \propto \dfrac{B}{B}$

$\propto 1$; that is, $\dfrac{A}{B}$ is constant.

(205.) *If one quantity vary as two others jointly, either of the latter varies as the first directly and the other inversely.*

Let $V \propto FT$, then by Art. 203,

$$F \propto \frac{V}{T} \text{ or } T \propto \frac{V}{F}.$$

(206.) Cor. If the product of two quantities be invariable, those quantities vary inversely as each other.

Let $B \times P$ be constant, or $B \times P \propto 1$; by division,

$$B \propto \frac{1}{P}.$$

(207.) *If*

(207.) *If four quantities be always proportionals, and one or two of them be invariable, we may find how the others vary.*

Ex. Let $p$, $q$, $r$, $s$, be always proportionals, and let $p$ be invariable, then $s \propto qr$. Because $ps = qr$ (Art. 172), $ps \propto qr$; and since $p$ is constant, $s \propto qr$, (Art. 199).

(208.) *If one quantity vary as a second, and a third as a fourth, the product of the first and third will vary as the product of the second and fourth.*

Let $A \propto B$ and $C \propto D$, then $AC \propto BD$.
Because $A : a :: B : b$
  and $C : c :: D : d$
    $AC : ac :: BD : bd$ (Art. 187);
that is, $AC \propto BD$.

(209.) *When the increase or decrease of one quantity depends upon the increase or decrease of two others, and it appears that if either of these latter be invariable, the first varies as the other; when they both vary, the first is as their product.*

Let $S \propto V$ when $T$ is given,
  and $S \propto T$ when $V$ is given;
when neither $T$ nor $V$ is given, $S \propto TV$. The variation of $S$ depends upon the variations of the two quantities $T$ and $V$; let the changes take place separately, and whilst $T$ is changed to $t$, let $S$ be changed to $S^{\scriptscriptstyle 1}$; then by the supposition, $S : S^{\scriptscriptstyle 1} :: T : t$; but this value $S^{\scriptscriptstyle 1}$ will again be changed to $s$, by the variation of $V$, and in the same proportion that $V$ is changed; that is, $S^{\scriptscriptstyle 1} : s :: V : v$, and by compounding this with the last proportion, $S^{\scriptscriptstyle 1} S : S^{\scriptscriptstyle 1} s :: TV : tv$; or, $S : s :: TV : tv$ (Art. 184).

(210.) In

(210.) In the same manner, if there be any number of magnitudes $P$, $Q$, $R$, $S$, each of which varies as another, $V$, when the rest are constant; when they are all changed, $V$ varies as their product.

## ON ARITHMETICAL PROGRESSION.

(211.) Quantities are said to be in *arithmetical progression*, when they increase or decrease by a common difference.

Thus, 1, 3, 5, 7, 9, &c. $a$, $a+b$, $a+2b$, $a+3b$, &c. $a$, $a-b$, $a-2b$, $a-3b$, &c. are in arithmetical progression.

Hence it is manifest, that if $a$ be the first term and $a+b$ the second, $a+2b$ is the third, $a+3b$ the fourth, &c. and $a+\overline{n-1}.b$ the $n^{th}$ term.

(212.) *The sum of a series of quantities in arithmetical progression is found by multiplying the sum of the first and last terms by half the number of terms.*

Let $a$ be the first term, $b$ the common difference, $n$ the number of terms, and $s$ the sum of the series: Then,

$$a \qquad +\overline{a+b} \qquad +\overline{a+2b}....a+\overline{n-1}.b = s,$$

or, $\overline{a+\overline{n-1}.b} + a+\overline{n-2}.b + a+\overline{n-3}.b ..... + a = s.$

Sum, $2a+\overline{n-1}.b + 2a+\overline{n-1}.b + 2a+\overline{n-1}.b + $ &c. to $n$ terms, $= 2s.$

or, $\overline{2a+\overline{n-1}.b} \times n = 2s.$

and $s = \overline{2a+\overline{n-1}.b} \times \dfrac{n}{2}.$

(213.) Cor.

(213.) Cor. Any three of the quantities $s$, $a$, $n$, $b$, being given, the fourth may be found from the equation $s = \overline{2a + n - 1.b} \times \dfrac{n}{2}$.

## Ex. 1.

To find the sum of 14 terms of the series 1, 3, 5, 7, &c.

Here $a = 1$, $b = 2$; $n = 14$; therefore, $s = \overline{2 + 26} \times 7 = 196$.

## Ex. 2.

Required the sum of 9 terms of the series 11, 9, 7, 5, &c.

In this case $a = 11$, $b = -2$, $n = 9$; therefore, $s = \overline{22 - 16} \times \dfrac{9}{2} = 6 \times \dfrac{9}{2} = 27$.

## Ex. 3.

If the first term of an arithmetical progression be 14, and the sum of 8 terms be 28, what is the common difference?

Since $\overline{2a + n - 1.b} \times \dfrac{n}{2} = s$,

$$2a + \overline{n - 1}.b = \dfrac{2s}{n}$$

$$\overline{n - 1}.b = \dfrac{2s}{n} - 2a = \dfrac{2s - 2an}{n};$$

therefore, $b = \dfrac{2s - 2an}{n.\overline{n - 1}}$. In the case proposed, $s = 28$, $a = 14$, $n = 8$; therefore, $b = \dfrac{56 - 224}{8 \times 7} = \dfrac{7 - 28}{7} = -3$.

Hence, the series is 14, 11, 8, 5, &c.

ON

## ON GEOMETRICAL PROGRESSION.

(214.) Quantities are said to be in *geometrical progression,* or continual proportion, when the first is to the second, as the second to the third, and as the third to the fourth, &c. that is, when every succeeding term is a certain multiple, or part of the preceding term.

If $a$ be the first term, and $ar$ the second, the series will be $a$, $ar$, $ar^2$, $ar^3$, $ar^4$, &c.

For $a : ar :: ar : ar^2 :: ar^2 : ar^3$, &c.

(215.) The constant multiplier is called the *common ratio,* and it may be found by dividing the second term by the first, or any other term by that which precedes it.

(216.) *If quantities be in geometrical progression, their differences are in geometrical progression.*

Let $a$, $ar$, $ar^2$, $ar^3$, $ar^4$, &c. be the quantities; their differences, $ar - a$, $ar^2 - ar$, $ar^3 - ar^2$, $ar^4 - ar^3$, &c. form a geometrical progression, whose first term is $ar - a$, and common ratio $r$.

(217.) *Quantities in geometrical progression are proportional to their differences.*

For $a : ar :: ar - a : ar^2 - ar :: ar^2 - ar : ar^3 - ar^2$, &c.

(218.) *In any geometrical progression, the first term is to the third, as the square of the first to the square of the second.*

Let $a$, $ar$, $ar^2$, &c. be the progression; then $a : ar^2 :: a^2 : a^2 r^2$.

Hence

Hence it appears, that the duplicate ratio of two quantities (Euc. Def. 10. 5), is the ratio of their squares.

(219.) In the same manner it may be shewn, that the first term, is to the $\overline{n+1}^{th}$ term, as the first raised to the $n^{th}$ power, to the second raised to the same power.

(220.) *If any terms be taken at equal intervals in a geometrical progression, they will be in geometrical progression.*

Let $a$, $ar$...$ar^n$.....$ar^{2n}$.....$ar^{3n}$.....&c. be the progression, then $a$, $ar^n$, $ar^{2n}$, $ar^{3n}$ &c. are at the interval of $n$ terms, and form a geometrical progression, whose common ratio is $r^n$.

(221.) *If the two extremes, and the number of terms in a geometrical progression be given, the means may be found.*

Let $a$ and $b$ be the extremes, $n$ the number of terms, and $r$ the common ratio; then the progression is $a$, $ar$, $ar^2$, $ar^3$.....$ar^{n-1}$; and since $b$ is the last term, $ar^{n-1}$ $=b$, and $r^{n-1}=\dfrac{b}{a}$; therefore $r=\overline{\dfrac{b}{a}}\Big|^{\frac{1}{n-1}}$ ; and $r$ being thus known, the terms of the progression $ar$, $ar^2$, $ar^3$, &c. are known.

(222.) *To find the sum of a series of quantities in geometrical progression, subtract the first term from the product of the last term and common ratio, and divide the remainder by the difference between the common ratio and unity.*

Let $a$ be the first term, $r$ the common ratio, $n$ the number

number of terms, $y$ the last term, and $s$ the sum of the series :

Then $a + ar + ar^2 \ldots + ar^{n-2} + ar^{n-1} = s$; and multiplying both sides by $r$,

$$ar + ar^2 + ar^3 \ldots + ar^{n-1} + ar^n = rs$$

Sub. $a + ar + ar^2 + ar^3 \ldots + ar^{n-1} \qquad = s$

Rem. $-a + ar^n = rs - s = \overline{r-1} \times s$

or, $s = \dfrac{ar^n - a}{r-1} = \dfrac{ry - a}{r-1}$.

(223.) Cor. 1. From the equation $s = \dfrac{ry - a}{r-1}$, any three of the quantities, $s$, $r$, $y$, $a$, being given, the fourth may be found.

(224.) Cor. 2. When $r$ is a proper fraction, as $n$ increases, the value of $r^n$, or of $ar^n$, decreases, and when $n$ is increased without limit, $ar^n$ becomes less, with respect to $a$, than any magnitude that can be assigned; and therefore $s = \dfrac{-a}{r-1} = \dfrac{a}{1-r}$.

This quantity $\dfrac{a}{1-r}$ which we call the *sum* of the series, is the *limit* to which the sum of the terms approaches, but never actually attains; it is however the true representative of the series continued *sine fine*; for this series arises from the division of $a$ by $1 - r$; and therefore $\dfrac{a}{1-r}$ may, without error, be substituted for it.

## Ex. 1.

To find the sum of 20 terms of the series, 1, 2, 4, 8, &c.

H                                                        Here

Here $a = 1$, $r = 2$, $n = 20$; therefore,

$$s = \frac{1 \times 2^{20} - 1}{2 - 1} = 2^{20} - 1.$$

### Ex. 2.

Required the sum of 12 terms of the series 64, 16, 4, &c.

Here $a = 64$, $r = \frac{1}{4}$, $n = 12$; therefore,

$$s = \frac{\frac{64}{4^{12}} - 64}{\frac{1}{4} - 1} = \frac{64 \times 4^{12} - 64}{4^{12} - 4^{11}} = \frac{64}{4^{11}} \times \frac{4^{12} - 1}{4 - 1}.$$

### Ex. 3.

Required the sum of 12 terms of the series $1, -3,$ $9, -27$, &c.

In this case, $a = 1$, $r = -3$, $n = 12$;

therefore, $s = \dfrac{\overline{-3}\,|^{12} - 1}{-3 - 1} = -\dfrac{3^{12} - 1}{4}$

### Ex. 4.

To find the sum of the series $1 - \frac{1}{2} + \frac{1}{4} - \frac{1}{8} + $&c. in infinitum.

Here $a = 1$, $r = -\frac{1}{2}$; therefore (Art. 224),

$$s = \frac{1}{1 + \frac{1}{2}} = \frac{2}{3}.$$

(225.) Recurring decimals are quantities in geometrical progression, where $\frac{1}{10}$, $\frac{1}{100}$, $\frac{1}{1000}$, &c. is the common ratio, according as one. two, three, &c.

figures

figures recur; and the vulgar fraction, corresponding to such a decimal, is found by summing the series.

### Ex. 5.

Required the vulgar fraction corresponding to the decimal 123 123 123 &c.

Let .123123123 &c. $= s$; then, as in Art. 222, multiply both sides by 1000; and 123.123123123 &c. $= 1000s$, and by subtracting the former equation from the latter, $123 = 999s$; therefore $s = \dfrac{123}{999} = \dfrac{41}{333}$.

## On PERMUTATIONS and COMBINATIONS.

(226.) The different orders in which any quantities can be arranged, are called their *permutations*.

Thus, the permutations of $a$, $b$, $c$, taken two and two together, are $ab$, $ba$, $ac$, $ca$, $bc$, $cb$.

(227.) The *combinations* of quantities are the different collections that can be formed out of them, without regarding the order in which the quantities are placed.

Thus, $ab$, $ac$, $bc$, are the combinations of the quantities $a$, $b$, $c$, taken two and two; $ab$ and $ba$, though different permutations, forming the same combination.

(228.) *The number of permutations that can be formed out of* n *quantities, taken two and two together, is* $n.\overline{n-1}$, *taken three and three together, is* $n.\overline{n-1}.\overline{n-2}$.

In $n$ things, $a$, $b$, $c$, $d$, &c. $a$ may be placed before each of the rest, and thus form $\overline{n-1}$ permutations; in

the

the same manner, there are $\overline{n-1}$ permutations in which $b$ stands first, and so of the rest; therefore there are, upon the whole, $n.\overline{n-1}$ permutations of this kind $ab$, $ba$, $ac$, $ca$, &c.

Again, of $n-1$ things $b$, $c$, $d$, &c. taken two and two together, there are $\overline{n-1}.\overline{n-2}$ permutations, by the former part of the article,· and by prefixing $a$ to each of these, there are $\overline{n-1}.\overline{n-2}$ permutations, taken three and three, in which $a$ stands first; the same may be said of $b$, $c$, $d$, &c. therefore there are, upon the whole, $n.\overline{n-1}.\overline{n-2}$ such permutations.

(229.) Cor. By following the same method, it appears, that in $n$ things, if $r$ of them be always taken together, there are $n.\overline{n-1}.\overline{n-2}.\overline{n-3}....\overline{n-r+1}$ permutations.

(230.) *The number of combinations that can be formed out of* n *things, taken two and two together, is* n. $\dfrac{n-1}{2}$; *taken three and three together, the number is* n. $\dfrac{n-1}{2}.\dfrac{n-2}{3}$.

The number of permutations in the first case is $n.\overline{n-1}$, but each combination $ab$, admits of two permutations, $ab$, $ba$; therefore there are twice as many permutations as combinations, or the number of combinations is $n.\dfrac{n-1}{2}$.

Again, there are $n.\overline{n-1}.\overline{n-2}$ permutations in $n$ things, taken three and three together; and each combination of three things admits of $3.2.1$ permutations (Art. 228); therefore there are $3.2.1$ times as many

many permutations as combinations, and consequently the number of combinations is $\dfrac{n.\overline{n-1}.\overline{n-2}}{1.2.3}$.

(231.) Cor. In the same manner it appears, that the number of combinations, in $n$ things, each of which contains $r$ of them, is $\dfrac{n.\overline{n-1}.\overline{n-2}...\overline{n-r+1}}{1.2.3........r}$.

## ON THE BINOMIAL THEOREM.

(232.) The method of raising a binomial to any power, by repeated multiplication, has before been laid down (Art. 117). The same thing may be done much more expeditiously by the following general rule, which is called the *Binomial Theorem*.

Let $x + a$ be the given binomial; and it's $n^{th}$ power is $x^n + nax^{n-1} + n.\dfrac{n-1}{2}a^2x^{n-2} + n.\dfrac{n-1}{2}\ \dfrac{n-2}{3}a^3x^{n-3} +$ &c. Where the index of $x$, beginning from $n$ is diminished by unity, and the index of $a$, beginning from 0, is increased by unity, in every succeeding term. Also, the coefficient of each term is found by multiplying the coefficient of the preceding term by the index of $x$ in that term, and dividing by the index of $a$ increased by unity.

Thus, $\overline{x+a}\,^6 = x^6 + 6ax^5 + \dfrac{6.5}{2}a^2x^4 + \dfrac{6.5.4}{2.3}a^3x^3 +$ $\dfrac{6.5.4.3}{2.3.4}a^4x^2 + \dfrac{6.5.4.3.2}{2.3.4.5}a^5x + \dfrac{6.5.4.3.2.1}{2.3.4.5.6}a^6 = x^6 + 6\,ax^5$ $+ 15\,a^2x^4 + 20\,a^3x^3 + 15\,a^4x^2 + 6\,a^5x + a^6.$

To

To investigate this theorem, suppose $n$ quantities, $x+a$, $x+b$, $x+c$, &c. multiplied together; it is manifest that the first term of the product will be $x^n$, and that, $x^{n-1}$, $x^{n-2}$, &c. the other powers of $x$, will all be found in the remaining terms, with different combinations of $a$, $b$, $c$, $d$, &c.

Let $\overline{x+b}.\overline{x+c}.\overline{x+d}$. &c. $=x^{n-1}+Px^{n-2}+Qx^{n-3}$ $+$&c. and $\overline{x+a}.\overline{x+b}.\overline{x+c}.\overline{x+d}$. &c. $=x^n+Ax^{n-1}$ $+Bx^{n-2}+$&c. then $x^n+Ax^{n-1}+Bx^{n-2}+$ &c. and $\overline{x+a}\times\overline{x^{n-1}+Px^{n-2}+Qx^{n-3}}+$&c. or,

$$\left.\begin{array}{l} x^n+Px^{n-1}+Qx^{n-2}+\&c. \\ \quad +ax^{n-1}+aPx^{n-2}+\&c. \end{array}\right\} \text{ are the same series;}$$

therefore $A=P+a$, $B=Q+aP$, &c. that is, by introducing one factor, $x+a$, into the product, the coefficient of the second term is increased by $a$, and by introducing $x+b$ into the product, that coefficient is increased by $b$, &c. therefore the whole value of $A$ is $a+b+c+d+$&c. Again, by the introduction of one factor, $x+a$, the coefficient of the third term, $Q$, is increased by $aP$, i. e. by $a$ multiplied by the preceding value of $A$, or by $a\times\overline{b+c+d+}$&c. and the same may be said with respect to the introduction of every other factor; therefore upon the whole,

$$B=a.\overline{b+c+d+}\&c.$$
$$+\,b.\overline{c+d+}\&c.$$
$$+\,c.\overline{d+}\&c.$$

In the same manner,

$$C=a.b.\overline{c+d+}\&c.$$
$$+a.c.\overline{d+}\&c.$$
$$+b.c.\overline{d+}\&c.$$

and so on; that is, $A$ is the sum of the quantities $a$, $b$, $c$,

$b$, $c$, &c. $B$ is the sum of the products of every two; $C$ is the sum of the products of every three, &c. &c.

Let $a=b=c=d=$&c. then $A$, or $a+b+c+d+$ &c. $=na$; $B=ab+ac+bc+$ &c. $=a^2 \times$ the number of combinations of $a$, $b$, $c$, $d$, &c. taken two and two,

$= n. \dfrac{n-1}{2} a^2$ (Art. 230); in the same-manner it appears, that $C = n.\dfrac{n-1}{2}.\dfrac{n-2}{3} a^3$, &c. And $\overline{x+a}.\overline{x+b}.$

$\overline{x+c}.$ &c. to $n$ factors $=\overline{x+a}\vert^n$; therefore $\overline{x+a}\vert^n = x^n$

$+nax^{n-1}+n.\dfrac{n-1}{2}a^2x^{n-2}+n.\dfrac{n-1}{2}.\dfrac{n-2}{3}a^3x^{n-3}+$&c.

This proof applies only to those cases in which $n$ is a whole positive number; but the rule holds when the index is fractional, or negative.

Let $\overline{1+x}\vert^{\frac{m}{n}} = 1+ax+bx^2+cx^3+$&c.

then $\overline{1+y}\vert^{\frac{m}{n}} = 1+ay+by^2+cy^3+$&c.

and, $\overline{1+x}\vert^{\frac{m}{n}} - \overline{1+y}\vert^{\frac{m}{n}} = a.\overline{x-y}+b.\overline{x^2-y^2}+c.\overline{x^3-y^3}$ $+$&c.

Assume $1+x=w^n$, and $1+y=v^n$, then $w^n-v^n=$ $x-y$, $\overline{1+a}\vert^{\frac{m}{n}}=w^m$, $\overline{1+y}\vert^{\frac{m}{n}}=v^m$, hence,

$w^m-v^m = a.\overline{x-y}+b.\overline{x^2-y^2}+c.\overline{x^3-y^3}+$&c.

and $\dfrac{w^m-v^m}{w^n-v^n} = a+b\dfrac{x^2-y^2}{x-y}+c.\dfrac{x^3-y^3}{x-y}+$ &c. $= a+$

$b.\overline{x+y}+c.\overline{x^2+xy+y^2}+$&c. (Art. 87. Ex. 5); or, dividing the numerator and denominator of the fraction $\dfrac{w^m-v^m}{w^n-v^n}$ by their common measure $w-v$, we have

$\dfrac{w^{m-1}+w^{m-2}v+w^{m-3}v^2.....+v^{m-1} \; (m)}{w^{n-1}+w^{n-2}v+w^{n-3}v^2.....+v^{n-1} \; (n)} = a+b.\overline{x+y}+$

$c.$

$\overline{c.x^2 + xy + y^2} + \&c.$ Now let $x = y$, then $w = v$, and

the equation becomes $\dfrac{m w^{m-1}}{n w^{n-1}} = \left(\dfrac{m w^m}{n w^n} = \right) a + 2 b x +$

$3 c x^2 + \&c.$

And multiplying by $w^n = 1 + x$,

$$\left.\begin{aligned} \frac{m}{n} . w^m &= a + 2 b x + 3 c x^2 + \&c. \\ &\quad + \ a x + 2 b x^2 + \&c. \end{aligned}\right\}$$

But $\dfrac{m}{n} . w^m = \dfrac{m}{n} . \overline{1 + x}^{\frac{m}{n}} = \dfrac{m}{n} . \overline{1 + a x + b x^2 + c x^3 + \&c.}$

therefore $\dfrac{m}{n} . \overline{1 + a x + b x^2 + c x^3 + \&c.} = a + 2 b x + 3 c x^2 + \&c.$

$$+ \ a x + 2 b x^2 + \&c.$$

And since these must form the same series, $\dfrac{m}{n} = a$;

$\dfrac{m}{n} . a = 2 b + a$, or $b = \dfrac{\overline{\dfrac{m}{n}} . \overline{\dfrac{m}{n} - 1}}{2}$; $\dfrac{m}{n} . b = 3 c + 2 b$, and $3 c$

$= \overline{\dfrac{m}{n} - 2} . b$, or $c = \dfrac{\overline{\dfrac{m}{n}} . \overline{\dfrac{m}{n} - 1} . \overline{\dfrac{m}{n} - 2}}{2.3}$; $\&c.$ Nearly in

the same manner, the proposition may be proved when

$\dfrac{m}{n}$ is negative.*

Ex. 1.

---

* See Encyclopædia Britan. vol. I. page 651.

It may not be improper to state, briefly, the nature of the proof alluded to in former Editions of this Work, as the principle is of extensive application.

It is usually taken for granted, and may without much difficulty be proved, that whether $r$ is positive or negative, whole or fractional, $\overline{1 + x}^r$ may properly be expressed by $1 + a x + b x^2 + c x^3 + \&c.$ where $a, b, c, \&c.$ are definite magnitudes, not dependent on

the

## Ex. 1.

$$\overline{a^2+x^2}\rvert^n = a^{2n}\times\overline{1+\frac{x^2}{a^2}}\rvert^n = a^{2n}.\overline{1+n.\frac{x^2}{a^2}+n.\frac{n-1}{2}.\frac{x^4}{a^4}+}$$

$$\&\text{c.} = a^{2n}+na^{2n-2}x^2+n.\frac{n-1}{2}.a^{2n-4}x^4+\&\text{c.}$$

## Ex. 2.

$$\overline{1+x}\rvert^{-n} = 1 - nx + n.\frac{n+1}{2}x^2 - n.\frac{n+1}{2}.\frac{n+2}{3}x^3+\&\text{c.}$$

## Ex. 3.

$$\overline{1+x}\rvert^{\frac{1}{n}} = 1+\frac{1}{n}x+\frac{1}{n}.\frac{\frac{1}{n}-1}{2}x^2+\frac{1}{n}.\frac{\frac{1}{n}-1}{2}.\frac{\frac{1}{n}-2}{3}x^3+\&\text{c.}$$

$$= 1+\frac{1}{n}x-\frac{n-1}{2n^2}x^2+\frac{\overline{n-1}.\overline{2n-1}}{2.3.n^3}x^3-\&\text{c.}$$

## Ex. 4.

$$\overline{1+x}\rvert^{-\frac{1}{n}} = 1-\frac{1}{n}x+\frac{n+1}{2n^2}x^2-\frac{\overline{n+1}.\overline{2n+1}}{2.3.n^3}x^3+\&\text{c.}$$

(233.) COR. 1. If either term of the binomial be
negative, it's odd powers will be negative (Art. 114);
and

---

the value of $x$. Let $a=r+z$, $b=r.\frac{r-1}{2}+v$, then, as appears from
what has been proved before (p. 118), when $r$ is any whole positive number, $a=r$, and $z=0$; that is, $z=0$ when $r$ is 1, 2, 3, &c.
or $z$ contains the factors $r-1$, $r-2$, $r-3$, &c. in inf. (Art. 269);
hence, $z=2.\overline{r-1}.\overline{r-2}.\overline{r-3}$. &c. in inf., which cannot be expressed in finite terms; consequently, $r+z$ cannot be expressed in
finite terms unless 2, that is, unless $z=0$; and since we know that
$a$, or $r+z$, may be expressed in finite terms, it follows that $z=0$,
and that $a=r$. In the same manner it appears that $b=r.\frac{r-1}{2}$, &c.

and consequently the signs of the terms, in which those odd powers are found, will be changed.

Ex. 5.

$$\overline{a^2 - x^2}\vert^n = a^{2n} - na^{2n-2}x^2 + n.\frac{n-1}{2}a^{2n-4}x^4 - \&c.$$

(234.) Cor. 2. If the index of the power, to which a binomial is to be raised, be a whole positive number, the series will terminate, because the coefficient $n.\dfrac{n-1}{2}$.

$\dfrac{n-2}{3}$. &c. will become nothing, when it is continued to $\overline{n+1}$ factors. In all other cases the number of terms will be indefinite.

(235.) Cor. 3. When the index is a whole positive number, the coefficients of the terms taken backward, from the end of the series, are respectively equal to the coefficients of the corresponding terms taken forward, from the beginning.

Thus, if $a+x$ be raised to the 8$^{\text{th}}$ power, the co-efficients are 1, 8, 28, 56, 70, 56, 28, 8, 1.

In general, the coefficient of the $\overline{n+1}^{\text{th}}$ term is $\dfrac{n.\overline{n-1}.\overline{n-2}\ldots\ldots3.\,2.\,1}{1.\,2.\,3\ldots\ldots\overline{n-2}.\overline{n-1}.n} = 1$.   The coefficient of the $n^{\text{th}}$ term is $\dfrac{n.\overline{n-1}.\overline{n-2}\ldots\ldots3.\,2.}{1.\ 2.\ 3\ldots.\overline{n-2}.\overline{n-1}} = n$; of the $\overline{n-1}^{\text{th}}$ term, $\dfrac{n.\overline{n-1}.\overline{n-2}\ldots.3.}{1.\ 2.\ 3.\ldots.\overline{n-2}} = \dfrac{n.\overline{n-1}}{1.2}$; &c.

(236.) Cor. 4. The sum of the coefficients $1 + n + n.\dfrac{n-1}{2} + \&c.$ is $2^n$.

For

For if $x = a = 1$, then $\overline{x+a}\rvert^n = \overline{1+1}\rvert^n = 2^n = 1 + n +$ $n \cdot \dfrac{n-1}{2} + \&c.$

(237.) Cor. 5. Since

$$\overline{x+a}\rvert^n = x^n + n a x^{n-1} + n \cdot \frac{n-1}{2} a^2 x^{n-2} + \&c.$$

and $\overline{x-a}\rvert^n = x^n - n a x^{n-1} + n \cdot \dfrac{n-1}{2} a^2 x^{n-2} - \&c.$

by addition, $\overline{x+a}\rvert^n + \overline{x-a}\rvert^n = 2.x^n + 2.n.\dfrac{n-1}{2}a^2x^{n-2} +$ &c.

or $\dfrac{\overline{x+a}\rvert^n + \overline{x-a}\rvert^n}{2} = x^n + n \cdot \dfrac{n-1}{2} a^2 x^{n-2} + \&c.$

By subtracting one series from the other, $\dfrac{\overline{x+a}\rvert^n - \overline{x-a}\rvert^n}{2}$

$$= n a x^{n-1} + n \cdot \frac{n-1}{2} \cdot \frac{n-2}{3} a^3 x^{n-3} + \&c.$$

(238.) The trinomial $a + b + c$ may be raised to any power by considering two terms as one factor, and proceeding as before.

Thus, $\overline{a+b+c}\rvert^n = a^n + n.\overline{b+c}.a^{n-1} + n.\dfrac{n-1}{2}.\overline{b+c}\rvert^2.$
$a^{n-2} + \&c.$ and the powers of $b+c$ may be determined by the binomial theorem.

The theorem by which a multinomial may be raised to any power, is given by Mr. Demoivre, Analyt. page 87.

## ON SURDS.

(239.) *A quantity may be reduced to the form of a given surd, by raising it to the power whose root the surd expresses, and affixing the radical sign.*

Thus,

Thus, $a = \sqrt{a^2} = \sqrt[3]{a^3}$, &c. and $a + x = \overline{a+x}\vert^{\frac{m}{m}}$. In the same manner, the form of any surd may be altered; thus, $\overline{a+x}\vert^{\frac{1}{2}} = \overline{a+x}\vert^{\frac{2}{4}} = \overline{a+x}\vert^{\frac{3}{6}}$ &c. The quantities are here raised to certain powers, and the roots of those powers are again taken, therefore the values of the quantities are not altered.

(240.) *If two surds have the same index, their product is found by taking the product of the quantities under the signs, and retaining the common index.*

Thus, $a^{\frac{1}{n}} \times b^{\frac{1}{n}} = \overline{a\,b}\vert^{\frac{1}{n}}$ (Art. 124); $\sqrt{2} \times \sqrt{3} = \sqrt{6}$; $\overline{a+b}\vert^{\frac{1}{2}} \times \overline{a-b}\vert^{\frac{1}{2}} = \overline{a^2-b^2}\vert^{\frac{1}{2}}$.

(241.) If the surds have coefficients, the product of these coefficients must be prefixed.

Thus, $a\sqrt{x} \times b\sqrt{y} = ab\sqrt{xy}$.

(242.) We must observe that $\sqrt{-a^2} \times \sqrt{-a^2}$, or the square of $\sqrt{-a^2}$, is $-a^2$, because it is that quantity whose square root is $\sqrt{-a^2}$.

Cor. Hence, $\overline{x-a+\sqrt{-b^2}} \times \overline{x-a-\sqrt{-b^2}} = x^2 - 2ax + a^2 + b^2$.

(243.) *If the indices of two surds have a common denominator, let the quantities be raised to the powers expressed by their respective numerators, and their product may be found as before.*

Ex. $2^{\frac{2}{3}} \times 3^{\frac{1}{3}} = 8^{\frac{1}{3}} \times 3^{\frac{1}{3}} = \overline{24}\vert^{\frac{1}{3}}$; also $\overline{a+x}\vert^{\frac{1}{3}} \times \overline{a-x}\vert^{\frac{2}{3}} = \overline{\overline{a+x}\,.\,\overline{a-x}\vert^3}\vert^{\frac{1}{3}}$.

(244.) *If the indices have not a common denominator, they may be transformed to others of the same value,*

*value, with a common denominator, and their product found as in* Art. 243.

Ex. $\overline{a^2-x^2}\,|^{\frac{1}{4}} \times \overline{a-x}\,|^{\frac{1}{2}} = \overline{a^2-x^2}\,|^{\frac{1}{4}} \times \overline{a-x}\,|^{\frac{2}{4}} =$

$\overline{a^2-x^2 \times \overline{a-x}\,|^{2}}\,|^{\frac{1}{4}}$ ; again, $2^{\frac{1}{2}} \times 3^{\frac{1}{3}} = 2^{\frac{3}{6}} \times 3^{\frac{2}{6}} = \overline{8 \times 9}\,|^{\frac{1}{6}}$.

(See Art. 239.)

(245.) *If two surds have the same rational quantity under the radical signs, their product is found by making the sum of the indices the index of that quantity.*

Thus, $a^{\frac{1}{n}} \times a^{\frac{1}{m}} = a^{\frac{m}{mn}} \times a^{\frac{n}{mn}}$ (Art. 239), $= a^{\frac{m+n}{mn}}$ ; (See Art. 124.)

Ex. 2. $\sqrt{2} \times \sqrt[3]{2} = 2^{\frac{3}{6}+\frac{2}{6}} = 2^{\frac{5}{6}}$.

(246.) *If the indices of two quantities have a common denominator, the quotient of one divided by the other is obtained by raising them, respectively, to the powers expressed by the numerators of their indices, and extracting that root of the quotient which is expressed by the common denominator.*

For, $\dfrac{a^{\frac{1}{n}}}{b^{\frac{1}{n}}} = \overline{\dfrac{a}{b}}\,|^{\frac{1}{n}}$ ; and $\dfrac{a^{\frac{m}{n}}}{b^{\frac{p}{n}}} = \overline{\dfrac{a^m}{b^p}}\,|^{\frac{1}{n}} = \overline{\dfrac{a^m}{b^p}}\,|^{\frac{1}{n}}$ (Art. 125.)

Ex. $4^{\frac{1}{3}} \div 2^{\frac{2}{3}} = \overline{\dfrac{4}{8}}\,|^{\frac{1}{3}} = \dfrac{1}{\sqrt{2}}$ ; $\overline{\dfrac{l}{q}}\,|^{\frac{1}{m}} \div \overline{\dfrac{r}{s}}\,|^{\frac{1}{m}} = \overline{\dfrac{p s^2}{q r^2}}\,|^{\frac{1}{m}}$.

(247.) *If the indices have not a common denominator, reduce them to others of the same value, with a common denominator, and proceed as before.*

Ex. $\overline{a^2-x^2}\,|^{\frac{1}{2}} \div \overline{a^3-x^3}\,|^{\frac{1}{3}} = \overline{a^2-x^2}\,|^{\frac{3}{} } \div \overline{a^3-x^3}\,| =$

$\dfrac{\overline{a^2-x^2}\,|^{3}}{\overline{a^3-x^3}\,|^{2}}\,|^{\frac{1}{6}}$

(248.) *If*

(248.) *If two surds have the same rational quantity under the radical signs, their quotient is obtained by making the difference of the indices, the index of that quantity.*

Thus, $a^{\frac{1}{n}}$ divided by $a^{\frac{1}{m}}$, or $a^{\frac{m}{mn}}$ divided by $a^{\frac{n}{mn}}$ (Art. 239), that is $\dfrac{a^{\frac{m}{mn}}}{a^{\frac{n}{mn}}}$, is equal to $a^{\frac{m-n}{mn}}$; because these quantities, raised to the power $mn$, produce equal results $\dfrac{a^m}{a^n}$ and $a^{m-n}$.

Ex. 2.   $2^{\frac{1}{2}} \div 2^{\frac{1}{3}} = 2^{\frac{3}{6}} \div 2^{\frac{2}{6}} = 2^{\frac{1}{6}}$.

(249.) *The coefficient of a surd may be introduced under the radical sign, by first reducing it to the form of the surd, and then multiplying as in Art. 240.*

Exs.  $a\sqrt{x} = \sqrt{a^2} \times \sqrt{x} = \sqrt{a^2 x}$; $ay^{\frac{3}{7}} = \overline{a^7 y^3}\,\rvert^{\frac{1}{7}}$

$x\sqrt{2a-x} = \sqrt{2ax^2-x^3}$; $a \times \overline{a-x}\rvert^{\frac{3}{7}} = \overline{a^2 \times \overline{a-x}\,\rvert^3}\rvert^{\frac{1}{2}}$;

$4\sqrt{2} = \sqrt{16 \times 2} = \sqrt{32}$.

(250.) *Conversely, any quantity may be made the coefficient of a surd, if every part under the sign be divided by this quantity raised to the power whose root the sign expresses.*

Thus, $\sqrt{a^2-ax} = a^{\frac{1}{2}} \times \sqrt{a-x}$; $\sqrt{a^3-a^2 x} = a\sqrt{a-x}$;

$\overline{a^2-x^2}\,\rvert^{\frac{1}{n}} = a^{\frac{2}{n}} \times \overline{1-\dfrac{x^2}{a^2}}\,\rvert^{\frac{1}{n}}$; $\sqrt{60} = \sqrt{4 \times 15} = 2\sqrt{15}$;

$\overline{\dfrac{1}{b^2}-\dfrac{1}{x^2}}\,\rvert^{\frac{1}{2}} = \dfrac{1}{b}\sqrt{1-\dfrac{b^2}{x^2}}$; $\sqrt{-x} = \sqrt{x} \times \sqrt{-1}$.

(251.) When

(251.) *When surds have the same irrational part, their sum or difference is found by affixing the sum or difference of their coefficients to that irrational part.*

Thus, $a\sqrt{x} \pm b\sqrt{x} = \overline{a \pm b} \cdot \sqrt{x}$; $\sqrt{300} \pm 5\sqrt{3} = 10\sqrt{3} \pm 5\sqrt{3} = 15\sqrt{3}$, or $5\sqrt{3}$.

(252.) *The square root of a quantity cannot be partly rational and partly a quadratic surd.*

If possible, let $\sqrt{n} = a + \sqrt{m}$; then by squaring both sides, $n = a^2 + 2a\sqrt{m} + m$, and $2a\sqrt{m} = n - a^2 - m$; therefore, $\sqrt{m} = \dfrac{n - a^2 - m}{2a}$, a rational quantity, which is contrary to the supposition.

(253.) *In any equation* $x + \sqrt{y} = a + \sqrt{b}$, *consisting of rational quantities and quadratic surds, the rational parts on each side are equal, and also the irrational parts.*

If $x$ be not equal to $a$, let $x = a + m$; then $a + m + \sqrt{y} = a + \sqrt{b}$, or $m + \sqrt{y} = \sqrt{b}$; that is, $\sqrt{b}$ is partly rational and partly a quadratic surd, which is impossible (Art. 252); therefore $x = a$, and consequently $\sqrt{y} = \sqrt{b}$.

(254.) *If two quadratic surds* $\sqrt{x}$ *and* $\sqrt{y}$, *cannot be reduced to others which have the same irrational part, their product is irrational.*

If possible, let $\sqrt{xy} = rx$, where $r$ is a whole number or a fraction. Then $xy = r^2x^2$, and $y = r^2x$; therefore $\sqrt{y} = r\sqrt{x}$; that is, $\sqrt{y}$ and $\sqrt{x}$ may be so reduced as to have the same irrational part, which is contrary to the supposition.

(255.) *One*

(255.) *One quadratic surd,* $\sqrt{x}$, *cannot be made up of two others,* $\sqrt{m}$ *and* $\sqrt{n}$, *which have not the same irrational part.*

If possible, let $\sqrt{x} = \sqrt{m} + \sqrt{n}$; then by squaring both sides, $x = m + n + 2\sqrt{mn}$, and $x - m - n = 2\sqrt{mn}$, a rational quantity equal to an irrational one (Art. 254); which is absurd.

(256.) *Let* $\overline{a+b}|^{\frac{1}{c}} = x + y$, *where c is an even number,* a *a rational quantity,* b *a quadratic surd,* x *and* y *one or both of them, quadratic surds, then* $\overline{a-b}|^{\frac{1}{c}}$ *$= x - y$.*

By involution, $a + b = x^c + cx^{c-1}y + c . \dfrac{c-1}{2} x^{c-2}y^2 +$

$c . \dfrac{c-1}{2} . \dfrac{c-2}{3} x^{c-3}y^3 + $ &c. and since $c$ is even, the odd terms of the series are rational, and the even terms irrational; therefore $a = x^c + c . \dfrac{c-1}{2} x^{c-2}y^2 + $ &c. and

$b = cx^{c-1}y + c . \dfrac{c-1}{2} . \dfrac{c-2}{3} x^{c-3}y^3 + $ &c. (Art. 253);

hence, $a - b = x^c - cx^{c-1}y + c.\dfrac{c-1}{2}x^{c-2}y^2 - c.\dfrac{c-1}{2}.\dfrac{c-2}{3}$

$x^{c-3}y^3 + $ &c. and consequently $\overline{a-b}|^{\frac{1}{c}} = x - y$.

(257.) *If* c *be an odd number,* a *and* b, *one or both quadratic surds, and* x *and* y *involve the same surds that* a *and* b *do, respectively, and also* $\overline{a+b}|^{\frac{1}{c}} = x + y$, *then* $\overline{a-b}|^{\frac{1}{c}} = x - y$.

By

By involution, $a+b=x^c+cx^{c-1}y+c.\dfrac{c-1}{2}x^{c-2}y^2+$

$c.\dfrac{c-1}{2}\dfrac{c-2}{3}x^{c-3}y^3+$&c. where the odd terms involve the same surd that $x$ does, because $c$ is an odd number, and the even terms, the same surd that $y$ does; and since no part of $a$ can consist of $y$, or it's parts (Art. 255), $a=x^c+c.\dfrac{c-1}{2}x^{c-2}y^2+$&c. and $b=$

$cx^{c-1}y+c.\dfrac{c-1}{2}.\dfrac{c-2}{3}x^{c-3}y^3+$&c. hence, $a-b=x^c-$

$cx^{c-1}y+c.\dfrac{c-1}{2}x^{c-2}y^2-c.\dfrac{c-1}{2}.\dfrac{c-2}{3}x^{c-3}y^3+$&c.

therefore, by evolution, $\overline{a-b}|^{\frac{1}{c}}=x-y$.

(258.) *The square root of a binomial, one of whose factors is a quadratic surd, and the other rational, may sometimes be expressed by a binomial, one or both of whose factors are quadratic surds.*

Let $a+\sqrt{b}$ be the given binomial, and suppose $\sqrt{a+\sqrt{b}}=x+y$, where $x$ and $y$ are one or both quadratic surds;

then $\sqrt{a-\sqrt{b}}=x-y$ (Art. 256),
by multiplication $\sqrt{a^2-b}=x^2-y^2$,
also, by squaring both sides of the first equation,
$$a+\sqrt{b}=x^2+2xy+y^2$$
and $a=x^2+y^2$ (Art. 253),
by addition, $a+\sqrt{a^2-b}=2x^2$.

by

by subtraction, $a - \sqrt{a^2 - b} = 2y^2$, and the root $x + y$

$$= \sqrt{\frac{a + \sqrt{a^2 - b}}{2}} + \sqrt{\frac{a - \sqrt{a^2 - b}}{2}}.$$

From this conclusion it appears, that the square root of $a + \sqrt{b}$ can only be expressed by a binomial of the form $x + y$, one or both of which are quadratic surds, when $a^2 - b$ is a perfect square.

The values of $x$ and $y$ are $\pm \sqrt{\dfrac{a + \sqrt{a^2 - b}}{2}}$ and $\pm \sqrt{\dfrac{a - \sqrt{a^2 - b}}{2}}$; there are therefore four different values of $x + y$; two of which were introduced in the operation (See Art. 149), and will not answer the conditions of the question.

## Ex. 1.

Required the square root of $3 + 2\sqrt{2}$.

In this case, $a = 3$, $\sqrt{b} = 2\sqrt{2}$, and $a^2 - b = 9 - 8 = 1$; hence, $x = \sqrt{\dfrac{3+1}{2}} = \sqrt{2}$, and $y = \sqrt{\dfrac{3-1}{2}} = 1$; therefore, $x + y = \sqrt{2} + 1$.

## Ex. 2.

Required the square root of $7 - 2\sqrt{10}$.

Here, $a = 7$, $\sqrt{b} = 2\sqrt{10}$, $a^2 - b = 9$; hence, $x = \sqrt{\dfrac{7+3}{2}} = \sqrt{5}$, and $y = \sqrt{\dfrac{7-3}{2}} = \sqrt{2}$; therefore, $x - y = \sqrt{5} - \sqrt{2}$, the root required.

Ex. 3.

### Ex. 3.

Required the square root of $4\sqrt{-5}-1$.

Here, $a=-1$, $\sqrt{b}=4\sqrt{-5}$, $a^2-b=81$; hence, $x=$

$\sqrt{\dfrac{-1+9}{2}}=2$, and $y=\sqrt{\dfrac{-10}{2}}=\sqrt{-5}$; and the

root required is $2+\sqrt{-5}$.

✝ (259.) *The $c^{th}$ root of a binomial, one or both of whose factors are possible quadratic surds, may sometimes be expressed by a binomial of that description.*

Let $A+B$ be the given binomial surd, in which both terms are possible; the quantities under the radical signs whole numbers; and $A$ is greater than $B$.

Let $\sqrt[c]{A+B} \times \sqrt{Q} = x+y$

then $\sqrt[c]{A-B} \times \sqrt{Q} = x-y$, (Art. 257);

by mult. $\sqrt[c]{A^2-B^2} \times Q = x^2-y^2$; let $Q$ be so assumed that $A^2-B^2 \times Q$ may be a perfect $c^{th}$ power $=n^c$, then $x^2-y^2=n$.

Again, by squaring both sides of the two first equations, we have

$$\sqrt[c]{A+B}^2 \times Q = x^2 + 2xy + y^2$$

$$\sqrt[c]{A-B}^2 \times Q = x^2 - 2xy + y^2$$

hence $\sqrt[c]{A+B}^2 \times Q + \sqrt[c]{A-B}^2 \times Q = 2x^2 + 2y^2$ which is always a whole number when the root is a binomial surd; take therefore $s$ and $t$ the nearest integer values of $\sqrt[c]{A+B}^2 \times Q$ and $\sqrt[c]{A-B}^2 \times Q$, one of which is greater, and the other less than the true value of the corresponding quantity; then since the sum of these surds is an integer, the fractional parts must destroy each other, and $2x^2 + 2y^2 = s+t$, exactly,

when

when the root of the proposed quantity can be obtained. We have therefore these two equations

$$x^2 - y^2 = n$$

$$x^2 + y^2 = \frac{s+t}{2}$$

therefore, $2x^2 = n + \dfrac{s+t}{2} = \dfrac{2n+s+t}{2}$

$$x^2 = \frac{2n+s+t}{4}$$

$$x = \frac{\sqrt{2n+s+t}}{2}$$

also, $2y^2 = \dfrac{s+t-2n}{2}$

and $y = \dfrac{\sqrt{s+t-2n}}{2}$;

therefore, if the root of the binomial $\sqrt[c]{\overline{A+B} \times \sqrt{Q}}$ be of the form $x+y$, it is $\dfrac{\sqrt{2n+s+t} + \sqrt{s+t-2n}}{2}$;

and the $c^{\text{th}}$ root of $A+B$ is $\dfrac{\sqrt{2n+s+t} + \sqrt{s+t-2n}}{2\sqrt[2c]{Q}}$.

## Ex. 1.

Required the cube root of $10 + \sqrt{108}$.

In this case, $A = \sqrt{108}$, $B = 10$; $A^2 - B^2 = 8$, and $8Q = n^3$; if therefore $Q = 1$, $n = 2$. Also,

$\sqrt[3]{\overline{A+B}|^2} = 7 + f$; and $\sqrt[3]{\overline{A-B}|^2} = 1 - f$, where $f$ is some fraction less than unity; therefore $s = 7$, $t = 1$; and $x + y = \dfrac{\sqrt{12} + 2}{2} = \sqrt{3} + 1$.

If therefore the cube root of $10 + \sqrt{108}$ can be expressed in the proposed form, it is $\sqrt{3} + 1$; which, on trial, is found to succeed.

Ex. 2.

## Ex. 2.

Let the cube root of $11 + 5\sqrt{7}$ be required.

Here, $A = 5\sqrt{7}$, $B = 11$, $A^2 - B^2 = 54$; therefore $54 Q = n^3$, and if $Q = 4$, $n^3 = 216$, and $n = 6$.

Also $\sqrt[3]{\overline{A + B}\vert^2 \times Q} = 13 + f$, $\sqrt[3]{\overline{A - B}\vert^2 \times Q} = 3 - f$; or $s = 13$, $t = 3$; hence, $x + y = \dfrac{\sqrt{28} + 2}{2} = \sqrt{7} + 1$,

and the quantity, to be tried for the root, is $\dfrac{\sqrt{7} + 1}{\sqrt{2}}$.

## Ex. 3.

To find the cube root of $2\sqrt{7} + 3\sqrt{3}$.

Here, $A = 2\sqrt{7}$, $B = 3\sqrt{3}$, $A^2 - B^2 = 1$; hence $Q = 1$, and $n = 1$. Also, $\sqrt[3]{\overline{A + B}\vert^2} = 4 + f$; $\sqrt[3]{\overline{A - B}\vert^2} = 1 - f$; or $s = 4$, $t = 1$; and $x + y = \dfrac{\sqrt{7} + \sqrt{3}}{2}$, the quantity to be tried for the root, which is found to succeed.

(260.) In the same manner, the $c^{th}$ root of $A - B$, is $\dfrac{\sqrt{2n + s + t} - \sqrt{s + t - 2n}}{2\sqrt[2c]{Q}}$; in which expression, when $A$ is less than $B$, $n$ is negative.

(261.) In the operation, it is required to find a number $Q$, such, that $\overline{A^2 - B^2} \times Q$ may be a perfect $c^{th}$ power; this will be the case, if $Q$ be taken equal to $\overline{A^2 - B^2}\vert^{c-1}$; but to find a less number which will answer this condition, let $A^2 - B^2$ be divisible by $a$, $a$, ..... $(m)$; $b$, $b$, .... $(n)$; $d$, $d$, ..... $(r)$; &c. in succession, that is, let $A^2 - B^2 = a^m \, b^n \, d^r$ &c. also let $Q =$

$Q = a^x\, b^y\, d^z$ &c. then $\overline{A^2 - B^2}$. $Q = a^{m+x} \times b^{n+y} \times d^{r+z}$ &c. which is a perfect $c^{th}$ power, if $x, y, z$, &c. be so assumed that $m+x$, $n+y$, $r+z$, are respectively equal to $c$, or some multiple of $c$. Thus, to find a number which multiplied by 180 will produce a perfect cube, divide 180 as often as possible by 2, 3, 5, &c. and it appears that $2.2.3.3.5 = 180$; if, therefore, it be multiplied by $2.3.5.5$, it becomes $2^3.3^3.5^3$, or $\overline{2.3.5}|^3$; a perfect cube.

If the index of the root to be extracted be an even number, the square root may be found by Art. 258, when it can be expressed by a binomial of the same description; and if half the index be an even number, the square root may again be taken, and so on, until the root remaining to be extracted is expressed by an odd number.

If $A$ and $B$ be divided by their greatest common measure, either integer or quadratic surd, in all cases where the $c^{th}$ root can be obtained by this method, $Q$ will either be unity, or some power of 2, less than $2^c$. See Dr. *Waring's Med. Alg. p.* 287.

THE END OF PART I.

THE

THE

# ELEMENTS OF ALGEBRA.

## PART II.

### ON THE NATURE OF EQUATIONS.

(262.) Any equation, involving the powers of one unknown quantity, may be reduced to the form $x^n - p x^{n-1} + q x^{n-2} -$ &c. $= 0$ ; where the whole expression is made equal to nothing, the terms are arranged according to the dimensions of the unknown quantity, the coefficient of the highest dimension is unity, and the coefficients, $p$, $q$, $r$, &c. are affected with their proper signs.

An equation, where the index of the highest power of the unknown quantity is $n$, is said to be of $n$ dimensions ; and in speaking simply of an equation of $n$ dimensions, we understand one reduced to the above form, unless the contrary be expressed.

(263.) Any quantity, $x^n - p x^{n-1} + q x^{n-2} ... + P x - Q$, may be supposed to arise from the multiplication of $\overline{x - a} . \overline{x - b} . \overline{x - c}$. &c. continued to $n$ factors.

For,

For, by actually multiplying the factors together, we obtain a quantity of $n$ dimensions, similar to the proposed quantity, $x^n - px^{n-1} + qx^{n-2} - $ &c. and if $a$, $b$, $c$, &c. can be so assumed that the coefficients of the corresponding terms in the two quantities become equal, the whole expressions coincide. And these coefficients may be made equal, because we shall have $n$ equations, to determine the $n$ quantities $a$, $b$, $c$, $d$, &c. (See Art. 145). If then the quantities, $a$, $b$, $c$, $d$, &c. be properly assumed, the equation $x^n - px^{n-1} + qx^{n-2} - $ &c. $= 0$, is the same with $\overline{x-a} . \overline{x-b} . \overline{x-c} .$ &c. $= 0 *$.

We cannot suppose $x^n - px^{n-1} + qx^{n-2} - $ &c. to be made up of more, or fewer, than $n$ simple factors; because, on either supposition, the result would not be of the same number of dimensions with the proposed quantity.

(264.) The quantities $a$, $b$, $c$, $d$, &c. are called *roots* of the equation, or *values* of $x$; because, if any one of them be substituted for $x$, the whole expression becomes nothing, which is the only condition proposed by the equation.

(265.) *If*

---

* This proof, which is usually given, is imperfect; for if the $n$ equations be reduced to one, containing only one of the quantities, $a$, this equation is $a^n - p a^{n-1} + q a^{n-2} - $ &c. $= 0$, which exactly coincides with the proposed equation; in supposing therefore that $a$ can be found, we take for granted the proposition to be proved. The subject has exercised the skill of the most eminent algebraical Writers, but their reasonings upon it are of too abstruse a nature to be introduced in this place : The Learner must, at present, take for granted, that an equation may be made up of as many simple factors as it has dimensions; and when he is farther advanced in the subject, he may consult Dr. Waring's *Algebra*, page 272; *Phil. Trans.* 1798, page 369; and 'Demonstratio Nova,' C. F. Gauss, Helmstadt, 1799.

(265.) *If the signs of the terms of an equation be all positive, it cannot have a positive root; and if the signs be alternately positive and negative, it cannot have a negative root.*

If $x^n + p x^{n-1} + q x^{n-2} + \&c. = 0$, and any positive quantity, $a$, be substituted for $x$, the result is positive; consequently $a$ is not a root of the equation.

If $x^n - p x^{n-1} + q x^{n-2} - \&c. = 0$, and a negative quantity, $-a$, be substituted for $x$, when $n$ is an odd number the result is negative, and when $n$ is an even number the result is positive; therefore $-a$ cannot, in either case, be a root of the equation.

(266.) *Every equation has as many roots as it has dimensions.*

If $x^n - p x^{n-1} + q x^{n-2} + \&c. = 0$, or $\overline{x-a}.\overline{x-b}.$ $\overline{x-c}.$ &c. to $n$ faetors $= 0$; there are $n$ quantities, $a$, $b$, $c$, &c. each of which, when substituted for $x$, makes the whole $= 0$, because in each case one of the factors becomes $= 0$; but any quantity different from these, as $e$, when substituted for $x$, gives the product $\overline{e-a}.\overline{e-b}.\overline{e-c}.$ &c. which does not vanish, because none of the factors vanish; that is, $e$ will not answer the condition which the equation requires.

(267.) When one of the roots, $a$, is obtained, the equation $\overline{x-a}.\overline{x-b}.\overline{x-c}.$ &c. $= 0$, or $x^n - p x^{n-1} + q x^{n-2} - \&c. = 0$, is divisible by $x-a$, without a remainder, and is thus reducible to $\overline{x-b}.\overline{x-c}.$ &c. $= 0$, an equation one dimension lower, whose roots are $b$, $c$, &c.

Ex. One root of the equation $y^3 + 1 = 0$, is $-1$, or $y + 1 = 0$, and the equation may be depressed to a quadratic, in the following manner:

$$y + 1$$

$$y+1)y^3 + 1 (y^2 - y + 1$$
$$\underline{y^3 + y^2}$$
$$-y^2$$
$$\underline{-y^2 - y}$$
$$+y+1$$
$$\underline{y+1}$$
$$*$$
$$\overline{\phantom{xxxxxx}}$$

Hence, the other two roots are the roots of the quadratic $y^2 - y + 1 = 0$

If two roots, $a$ and $b$, be obtained, the equation is divisible by $\overline{x - a} . \overline{x - b}$; and thus it may be reduced two dimensions lower.

Ex. Two roots of the equation $x^6 - 1 = 0$, are $+1$ and $-1$, or $x - 1 = 0$, and $x + 1 = 0$; therefore it may be depressed to a biquadratic by dividing by $\overline{x - 1}$. $\overline{x + 1}$, or by $x^2 - 1$.

$$x^2 - 1) x^6 - 1 (x^4 + x^2 + 1$$
$$\underline{x^6 - x^4}$$
$$+x^4$$
$$\underline{x^4 - x^2}$$
$$+x^2 - 1$$
$$\underline{x^2 - 1}$$
$$*$$
$$\overline{\phantom{xxxxxx}}$$

Hence, the equation $x^4 + x^2 + 1 = 0$, contains the other four roots. of the proposed equation.

(268.) Conversely, if the equation be divisible by $x - a$, without remainder, $a$ is a root; if by $\overline{x - a}$. $\overline{x - b}$,

$\overline{x-b}$, $a$ and $b$ are both roots; &c. Let $Q$ be the quotient arising from the division, then the equation is $\overline{x-a}.\overline{x-b}.Q=0$, in which, if $a$ or $b$ be substituted for $x$, the whole vanishes.

(269.) Cor. 1. If $a$, $b$, $c$, &c. be the roots of an equation, that equation is $\overline{x-a}.\overline{x-b}.\overline{x-c}.$ &c.$=0$. Thus, the equation whose roots are 1, 2, 3, 4, is $\overline{x-1}.\overline{x-2}.\overline{x-3}.\overline{x-4}=0$; or $x^4-10x^3+35x^2-50x+24=0$. The equation whose roots are 1, 2, and $-3$, is $\overline{x-1}.\overline{x-2}.\overline{x+3}=0$; or, $x^3-7x+6=0$.

(270.) Cor. 2. If the last term of an equation vanish, it is of the form $x^n-px^{n-1}+qx^{n-2}\ldots\ldots Px=0$, which is divisible by $x$, or $x-o$, without remainder; therefore, $o$ is one of it's roots; if the two last terms vanish, it is divisible by $x^2$, without remainder, or by $\overline{x-o}.\overline{x-o}$, that is, two of it's roots are $o$; &c.

(271.) *The coefficient of the second term of an equation, with it's proper sign, is the sum of the roots with their signs changed; the coefficient of the third term, is the sum of the products of every two roots with their signs changed; the coefficient of the fourth term, is the sum of the products of every three roots, with their signs changed, &c. and the last term is the product of all the roots, with their signs changed.*

Let $a$, $b$, $c$, &c. be the roots of the equation; then $\overline{x-a}.\overline{x-b}.\overline{x-c}.$ &c.$=0$, is that equation; and by Art. 232, it appears, that when these factors are multiplied together, the coefficient of the second term is the sum of the quantities $-a$, $-b$, $-c$; &c. the coefficient of the third term, the sum of the products of every two, &c. and the last term, which does

does not contain $x$, is the product of all those quantities.

(272.) Cor. 1. If the roots be all positive, the signs of the terms will be alternately $+$ and $-$. For the product of an odd number of negative quantities is negative, and of an even number, positive. But if the roots be all negative, the signs of all the terms will be positive, because the equation arises from the multiplication of the positive quantities $\overline{x+a}.\overline{x+b}.\overline{x+c}$. &c.

(273.) Cor. 2. Let the roots of the equation $x^n - px^{n-1}\ldots + Px^2 - Qx + R = 0$, be $a$, $b$, $c$, $d$, &c. then $R = abcd\ (n)$; $Q = bcd\ (n-1) + acd\ (n-1) + abd\ (n-1) +$ &c. and $\dfrac{Q}{R} = \dfrac{1}{a} + \dfrac{1}{b} + \dfrac{1}{c} +$ &c. that is, the coefficient of the last term but one, divided by the last term, is the sum of the reciprocals of the roots. In the same manner, $\dfrac{P}{R} = \dfrac{1}{ab} + \dfrac{1}{ac} + \dfrac{1}{ad} + \dfrac{1}{bc} + \dfrac{1}{bd} + \dfrac{1}{cd} +$ &c.

(274.) Any equation, it has been observed, may be conceived to arise from the multiplication of the simple factors $\overline{x-a}.\overline{x-b}.\overline{x-c}$. &c. or by taking two or more of these together, it may be supposed to arise from the multiplication of quadratic, cubic, &c. factors, if the dimensions of these factors, together, make up the dimensions of the proposed equation.

Thus, a cubic equation may be supposed to be the product of three simple factors, $\overline{x-a}.\overline{x-b}.\overline{x-c} = 0$; or of a quadratic and a simple factor, $\overline{x^2-px+q} \times \overline{x-c} = 0$.

(275.) *If*

(275.) *If the coefficients, in any equation, be whole numbers, the equation cannot have a fractional root.*

If possible, let $\frac{a}{b}$, a fraction in it's lowest terms, be a root of the equation $x^n - p x^{n-1} + q x^{n-2} -$ &c. $= 0$; then $\frac{a^n}{b^n} - \frac{p a^{n-1}}{b^{n-1}} + \frac{q a^{n-2}}{b^{n-2}} -$ &c. $= 0$; and by transposition, $\frac{a^n}{b^n} = \frac{p a^{n-1}}{b^{n-1}} - \frac{q a^{n-2}}{b^{n-2}} +$ &c. or $\frac{a^n}{b} = p a^{n-1} -$ $q a^{n-2} b +$ &c. that is, $\frac{a^n}{b}$, a fraction in it's lowest terms (Art. 190. Cor. 2), is equal to a whole number, which is absurd ; therefore $\frac{a}{b}$ is not a root of the equation.

(276.) The roots $a$, $b$, $c$, &c. of an equation are *impossible*, when, as is frequently the case, they involve the square root of a negative quantity.

(277.) *Impossible roots enter equations by pairs.*

If $a + \sqrt{-b^2}$ be a root of the equation $x^n - p x^{n-1}$ $+$ &c. $= 0$, then $a - \sqrt{-b^2}$ is also a root.

In the equation, for $x$ substitute $a + \sqrt{-b^2}$, and the result will consist of two parts, possible quantities, which involve the powers of $a$ and the even powers of $\sqrt{-b^2}$, and impossible quantities, which involve the odd powers of $\sqrt{-b^2}$; call the sum of the possible quantities $A$, and of the impossible $B$, then $A + B$ is the whole result. Let now, $a - \sqrt{-b^2}$, be substituted for $x$, and the possible part of the result will be the same as before, and the impossible part which arises from

from the odd powers of $-\sqrt{-b^2}$, will only differ from the former impossible part in it's sign; therefore the result is $A - B$; and since by the supposition $a + \sqrt{-b^2}$ is a root of the equation, $A + B = 0$; in which, as no part of $A$ can be destroyed by $B$, $A = 0$ and $B = 0$; therefore $A - B = 0$, that is, the result, arising from the substitution of $a - \sqrt{-b^2}$ for $x$, is nothing; or $a - \sqrt{-b^2}$ is a root of the equation.

The truth of the proposition is also manifest from this consideration, that if $x^2 - mx + n = 0$ be a quadratic factor of the equation (Art. 274), two values of $x$ are $\dfrac{m + \sqrt{m^2 - 4n}}{2}$, and $\dfrac{m - \sqrt{m^2 - 4n}}{2}$, which are either both possible, or both impossible.

(278.) COR. 1. Hence it follows, that an equation of an odd number of dimensions, must have, at least, one possible root; unless some of the coefficients are impossible, in which case the equation may have an odd number of impossible roots.

(279.) COR. 2. By the same mode of reasoning it appears that, when the coefficients are rational, surd roots of the form $\pm \sqrt{b}$, or $a \pm \sqrt{b}$, enter equations by pairs.

# ON THE TRANSFORMATION OF EQUATIONS.

(280.) *If the signs of all the terms in an equation be changed, it's roots are not altered.*

Let

Let $\overline{x-a}.\overline{x-b}.\overline{x-c}$. &c.$=0$; then $-\overline{x-a}.\overline{x-b}$. $\overline{x-c}$. &c. vanishes when $a$, $b$, $c$, &c. are substituted for $x$.

(281.) *If the signs of the alternate terms, beginning with the second, be changed, the signs of all the roots are changed.*

Let $x^n - px^{n-1} - qx^{n-2} +$&c. $= 0$, be an equation whose roots are $a$, $b$, $-c$, &c. for $x$ substitute $-y$, and, when $n$ is an even number, $y^n + py^{n-1} - qy^{n-2} -$ &c. $=0$; but when $n$ is an odd number, $-y^n - py^{n-1} + qy^{n-2} +$&c.$=0$, or changing all the signs (Art. 280), $y^n + py^{n-1} - qy^{n-2} -$ &c. $= 0$, as before; and since $x = -y$, or $y = -x$, the values of $y$ are $-a, -b, +c$, &c.

Ex. Let it be required to change the signs of the roots of the equation $x^3 - qx + r = 0$.

This equation with all it's terms is $x^3 + o - qx + r = 0$; and changing the signs of the alternate terms, we have $x^3 - o - qx - r = 0$, or $x^3 - qx - r = 0$, an equation whose roots differ from the roots of the former, only in their signs.

(282.) *To transform an equation into one whose roots are greater, or less than the corresponding roots of the original equation, by any given quantity.*

Let the roots of the equation $x^3 - px^2 + qx - r = 0$, be $a$, $b$, $c$; to transform it into one whose roots are $a+e$, $b+e$, $c+e$.

Assume $x+e=y$, or $x=y-e$; then,
$$\left.\begin{array}{l} x^3 = y^3 - 3\,ey^2 + 3\,e^2y - e^3 \\ -px^2 = \quad -py^2 + 2pey - pe^2 \\ +qx = \qquad\quad + qy - qe \\ -r = \qquad\qquad\quad -r \end{array}\right\} = 0.$$

In

In this last equation, since $y = x + e$, the values of $y$ are $a+e$, $b+e$, $c+e$.

If $y+e$ be substituted for $x$, the values of $y$ in the resulting equation will be $a-e$, $b-e$, $c-e$.

(283.) In general, let the roots of the equation $x^n - px^{n-1} + qx^{n-2} - \&c. = 0$, be $a$, $b$, $c$, &c. Assume $y = x - e$, or $x = y + e$, and by substitution,

$$\left.\begin{array}{l} x^n = y^n + ney^{n-1} + n.\dfrac{n-1}{2}\,e^2 y^{n-2}\ldots\ldots + ne^{n-1}y + e^n \\[2mm] -px^{n-1} = \quad -py^{n-1} - \overline{n-1}.pey^{n-2}\ldots - \overline{n-1}.pe^{n-2}y - pe^{n-1} \\[2mm] +qx^{n-2} = \qquad + \qquad qy^{n-2}\ldots + \overline{n-2}.qe^{n-3}y + qe^{n-2} \\[2mm] \&c. = \qquad\qquad \&c. \end{array}\right\} = 0,$$

and since $y = x - e$, the values of $y$, in this equation, are $a-e$, $b-e$, $c-e$, &c.

We may observe, that the last term of the transformed equation, $e^n - pe^{n-1} + qe^{n-2} - \&c.$ is the original quantity, with $e$ in the place of $x$; the coefficient of the last term but one, with it's proper sign, is obtained by multiplying every term of $e^n - pe^{n-1} + qe^{n-2} - \&c.$ by the index of $e$ in that term, and diminishing the index by unity; the coefficient of the last term but two,

$$n.\frac{n-1}{2}e^{n-2} - \overline{n-1}.\frac{n-2}{2}pe^{n-3} + \overline{n-2}.\frac{n-3}{2}qe^{n-4}\&c.\text{ is}$$

obtained in the same manner from the coefficient of the last term but one, dividing every term by 2; &c.

(284.) One use of this transformation is, to take away any term out of an equation. Thus, to transform an equation into one which shall want the *second* term, $e$ must be so assumed that $ne - p = 0$, or $e = \dfrac{p}{n}$

(where $p$ is the coefficient of the second term

with

with it's sign changed, and $n$ the index of the highest power of the unknown quantity); and if the roots of the transformed equation can be found, the roots of the original equation may also be found, because

$$x = y + \frac{p}{n}.$$

Ex. 1. Let $x^2 - px + q = 0$ be the proposed equation. Assume $x = y + \frac{p}{2}$, then

$$\left.\begin{array}{l} x^2 = y^2 + py + \dfrac{p^2}{4} \\[2mm] \quad - \quad -py - \dfrac{p^2}{2} \\[2mm] + \; q = \qquad + q \end{array}\right\} = 0,$$

or $y^2 - \dfrac{p^2}{4} + q = 0$; hence, $y^2 = \dfrac{p^2}{4} - q$, and $y = \pm$ $\sqrt{\dfrac{p^2}{4} - q}$; therefore $x = \dfrac{p}{2} \pm \sqrt{\dfrac{p^2}{4} - q}$.

Ex. 2. To transform the equation $x^3 - 9x^2 + 7x + 12 = 0$ into one which shall want the second term.

Assume $x = y + 3$, then

$$\left.\begin{array}{l} x^3 = y^3 + 9y^2 + 27y + 27 \\ -9x^2 = \quad -9y^2 - 54y - 81 \\ +7x = \qquad + 7y + 21 \\ +12 = \qquad\qquad +12 \end{array}\right\} = 0,$$

that is, $y^3 - 20y - 21 = 0$; and if the values of $y$ be $a$, $b$, $c$, the values of $x$ are $a+3$, $b+3$ and $c+3$.

(285.) To take away the *third* term of the equation, $e$ must be so assumed, that $n.\dfrac{n-1}{2}e^2 - \overline{n-1}.pe + q = 0.$

K                                                    In

In this case we shall have a quadratic to solve ; and in general, to take out the $m^{th}$ term, by this method, it will be necessary to solve an equation of $m-1$ dimensions.

Ex. To transform the equation $x^3 - 6x^2 + 9x - 1 = 0$ into one which shall want the third term.

Here $n=3$, $p=6$ and $q=9$ ; therefore $n \cdot \dfrac{n-1}{2} e^2 - \overline{n-1} \cdot pe + q = 0$ becomes $3e^2 - 12e + 9 = 0$, or $e^2 - 4e + 3 = 0$, in which the values of $e$ are 1 and 3. Let $x = y + 3$, then

$$\left.\begin{array}{lr} x^3 = y^3 + 9y^2 + 27y + 27 \\ -6x^2 = \quad\quad -6y^2 - 36y - 54 \\ +9x \quad\quad\quad\quad\quad + 9y + 27 \\ -1 = \quad\quad\quad\quad\quad\quad - 1 \end{array}\right\} = 0,$$

that is, $y^3 + 3y^2 - 1 = 0$. In the same manner, if $x = y + 1$, the transformed equation will want the third term.

(286.) *To transform an equation into one whose roots are the roots of the original equation multiplied by any given quantity.*

Let the roots of the equation $x^n - px^{n-1} + qx^{n-2} - $ &c. $= 0$, be $a$, $b$, $c$, &c. to transform it into one whose roots are $ma$, $mb$, $mc$, &c.

Assume $y = mx$, or $x = \dfrac{y}{m}$ ; then substitute this value for $x$, and the equation becomes $\dfrac{y^n}{m^n} - \dfrac{py^{n-1}}{m^{n-1}} + \dfrac{qy^{n-2}}{m^{n-2}}$ $-$ &c. $= 0$, or multiplying by $m^n$, $y^n - mpy^{n-1} + m^2qy^{n-2} - $ &c. $= 0$, an equation whose roots are $ma$, $mb$, $mc$, &c.

This

This equation differs from the former, only in having the successive terms multiplied by 1, $m$, $m^2$, $m^3$, &c.

(287.) Cor. 1. By this transformation, an equation may be cleared of fractions; or if the first term be affected with a coefficient, that coefficient may be taken away.

Ex. 1. Let $x^n - \dfrac{p\,x^{n-1}}{m} + \dfrac{q\,x^{n-2}}{n} - r x^{n-3} + \&c. = 0$;

by multiplication, $mn x^n - np x^{n-1} + mq x^{n-2} - mnr x^{n-3} + \&c. = 0$; transform this equation into one whose roots are $mn$ times as great, and $mny^n - mn^2 p y^{n-1} + m^3 n^2 q y^{n-2} - m^4 n^4 r y^{n-3} + \&c. = 0$, or $y^n - npy^{n-1} + m^2 n q y^{n-2} - m^3 n^3 r y^{n-3} + \&c. = 0$, an equation of the usual form.

Ex. 2. Let it be required to transform the equation $3y^3 - qy + r = 0$ into one in which the coefficient of the highest term is unity.

The equation with all it's terms is $3y^3 + o - qy + r = 0$; transform it into one whose roots are three times as great, by substituting $\dfrac{z}{3}$ for $y$; then $3z^3 + 3 \times o - 9qz + 27 r = 0$, or $z^3 - 3qz + 9r = 0$, an equation of the required form.

(288.) Cor. 2. In any equation, if the coefficients of the second, third, fourth terms, &c. be divisible, respectively, by $m$, $m^2$, $m^3$, &c. the roots have a common measure, $m$.

(289.) Cor. 3. An equation may be transformed into one whose roots are $\dfrac{1}{m}$ parts of the roots of the

K 2                                                        former,

former, by dividing the second, third, fourth, &c. terms by $m$, $m^2$, $m^3$, &c. respectively.

(290.) *To transform an equation into one whose roots are the reciprocals of the roots of the given equation.*

Let the roots of the equation $x^n - px^{n-1} + qx^{n-2}$ .... $- Px + Q = 0$, be $a$, $b$, $c$, &c. to transform it into one whose roots are $\dfrac{1}{a}$, $\dfrac{1}{b}$, $\dfrac{1}{c}$, &c.

Assume $y = \dfrac{1}{x}$, or $x = \dfrac{1}{y}$; then, by substitution, $\dfrac{1}{y^n}$

$- \dfrac{p}{y^{n-1}} + \dfrac{q}{y^{n-2}}$ .... $- \dfrac{P}{y} + Q = 0$, and multiplying by

$y^n$, $1 - py + qy^2 \ldots - Py^{n-1} + Qy^n = 0$; that is, $Qy^n -$

$Py^{n-1}$ .... $+ qy^2 - py + 1 = 0$; and since $y = \dfrac{1}{x}$, the

values of $y$ are $\dfrac{1}{a}$, $\dfrac{1}{b}$, $\dfrac{1}{c}$, &c.

(291.) Cor. 1. If any term in the given equation be wanting, the corresponding term will be wanting in the transformed equation; thus, if the original equation want the second term, the transformed equation will want the last term but one, &c. because the coefficients in the transformed equation are the coefficients in the original equation, in an inverted order.

(292.) Cor. 2. If the coefficients of the terms, taken from the beginning of an equation, be the same with the coefficients of the corresponding terms, taken from the end, with the same signs, the transformed will coincide with the original equation, and their roots will therefore be the same.

Let

Let $a$, $b$, $c$, be roots of the equation $x^n - px^{n-1} + qx^{n-2} \ldots \ldots + qx^2 - px + 1 = 0$; the transformed equation will be $y^n - py^{n-1} + qy^{n-2} \ldots \ldots + qy^2 - py + 1 = 0$, and $a$, $b$, $c$, must also be roots of this equation; but the roots of this equation are the reciprocals of the roots of the original equation, therefore $\frac{1}{a}$, $\frac{1}{b}$, $\frac{1}{c}$, are also roots of the original equation.

Ex. The roots of the equations, $x^4 - px^3 + qx^2 - px + 1 = 0$; $x^4 + qx^2 + 1 = 0$; and $x^4 + 1 = 0$, are of the form $a$, $b$, $\frac{1}{a}$, $\frac{1}{b}$.

(293.) Cor. 3. If the equation be of an odd number of dimensions, or if the middle term of an equation of an even number of dimensions be wanting, the same thing will hold when the signs of the corresponding terms, taken from the beginning and end, are different.

Ex. The roots of the equation $x^3 - px^2 + px - 1 = 0$, are of the form $1$, $a$, $\frac{1}{a}$. For in this case, if the signs of all the terms of the transformed equation be changed, it will coincide with the original equation; and by changing the signs of all the terms, we do not alter the roots. —(Art. 280).

(294.) The equations described in the two last corollaries are called *recurring* equations.

(295.) Cor. 4. One root of a recurring equation of an odd number of dimensions, will be $+ 1$, or $- 1$, according

according as the sign of the last term is $-$ or $+$ ; and the rest will be of the form $a$, $\dfrac{1}{a}$, $b$, $\dfrac{1}{b}$; &c.

For if $+1$, in the former case, and $-1$, in the latter, be substituted for the unknown quantity, the whole vanishes; thus, if $x^5 - px^4 + qx^3 - qx^2 + px - 1 = 0$, and for $x$ we substitute $+1$, it becomes $1 - p + q - q + p - 1 = 0$; and it appears from Art. 290, that if $a$, $b$, $c$, &c. be roots of the equation, $\dfrac{1}{a}$, $\dfrac{1}{b}$, $\dfrac{1}{c}$, &c. are also roots.

(296.) *To transform an equation into one whose roots are the squares of the roots of the given equation.*

Let $x^n - px^{n-1} + qx^{n-2} - rx^{n-3} + sx^{n-4} - $ &c. $= 0$; by transposition, $x^n + qx^{n-2} + sx^{n-4} + $ &c. $= px^{n-1} + rx^{n-3} + $ &c.; and by squaring both sides, $x^{2n} + 2qx^{2n-2} + \overline{q^2 + 2s}.x^{2n-4} + $ &c. $= p^2x^{2n-2} + 2prx^{2n-4} + $ &c. and again by transp. $x^{2n} - \overline{p^2 - 2q}.x^{2n-2} + \overline{q^2 - 2pr + 2s}.x^{2n-4} - $ &c. $= 0$; assume $y = x^2$, then $y^n - \overline{p^2 - 2q}.y^{n-1} + \overline{q^2 - 2pr + 2s}.y^{n-2} - $ &c. $= 0$, in which equation, the values of $y$ are the squares of the values of $x$.

(297.) Cor. If the roots of the original equation be $a$, $b$, $c$, &c. then $p^2 - 2q = a^2 + b^2 + c^2 + $ &c. $q^2 - 2pr + 2s = a^2b^2 + a^2c^2 + $ &c. (Art. 271).

Other transformations may be seen in Dr. Waring's *Meditationes Algebraicæ*, Prob. 5 ; and indeed, whoever would fully understand the nature of equations, must have recourse to that Work.

ON

# ON THE LIMITS OF THE ROOTS OF EQUATIONS.

(298.) If $a$, $b$, $c$, $-d$, &c. be the roots of an equation, taken in order, that is, $a$ greater than $b$, $b$ greater than $c$, &c.* the equation is $\overline{x-a}.\overline{x-b}.\overline{x-c}.\overline{x+d}.$&c. $=0$; in which, if a quantity greater than $a$ be substituted for $x$, as every factor is, on this supposition, positive, the result will be positive; if a quantity less than $a$, but greater than $b$, be substituted, the result will be negative, because the first factor will be negative and the rest positive. If a quantity between $b$ and $c$ be substituted, the result will again be positive, because the two first factors are negative and the rest positive, and so on. Thus, quantities which are limits to the roots of an equation, or between which the roots lie, if substituted, successively, for the unknown quantity, give results alternately positive and negative.

(299.) Conversely, if two magnitudes, when substituted for the unknown quantity, give results affected with different signs, an odd number of roots must lie between them; and if a series of magnitudes, taken in order, can be found, which give as many results, alternately positive and negative, as the equation has dimensions, these must be limits to the roots of the equation; because an odd number of roots lies between each

---

* In this series, the greater $d$ is, the less is $-d$. And whenever $a$, $b$, $c$, $-d$, &c. are said to be the roots of an equation taken in order, $a$ is supposed to be the greatest. Also, in speaking of the limits of the roots of an equation, we understand the limits of the possible roots.

each two succeeding terms of the series, and there are as many terms as the equation has dimensions; therefore this odd number cannot exceed unity.

(300.) If the results arising from the substitution of two magnitudes, for the unknown quantity, be both positive or both negative, either no root of the equation, or an even number of roots, lies between them.

(301.) Cor. If $m$, and every quantity greater than $m$, when substituted for the unknown quantity, give positive results, $m$ is greater than the greatest root of the equation.

(302.) *To find a limit greater than the greatest root of an equation.*

Let the roots of the equation be $a$, $b$, $c$, &c. transform it into one whose roots are $a-e$, $b-e$, $c-e$, &c. and if, by trial, such a value of $e$ be found, that every term of the transformed equation is positive, all it's roots are negative (Art. 265), and consequently $e$ is greater than the greatest root of the proposed equation.

Ex. 1. To find a number greater than the greatest root of the equation $x^3 - 5x^2 + 7x - 1 = 0$.

Assume $x = y + e$, and we have

$$\left.\begin{array}{c} y^3 + 3ey^2 + 3e^2y + e^3 \\ -5y^2 - 10ey - 5e^2 \\ + 7y + 7e \\ -1 \end{array}\right\} = 0,$$

in which equation, if 3 be substituted for $e$, each of the quantities, $e^3 - 5e^2 + 7e - 1$, $3e^2 - 10e + 7$, $3e - 5$, is positive, or all the values of $y$ are negative; therefore 3 is greater than the greatest value of $x$.

Ex. 2.

**Ex. 2.** In any cubic equation of this form, $x^3 - qx + r = 0$, $\sqrt{q}$ is greater than the greatest root.

By transforming the equation, as before,

$$\left. \begin{array}{r} y^3 + 3ey^2 + 3e^2y + e^3 \\ -\ qy - qe \\ +\ r \end{array} \right\} = 0,$$

and substituting $\sqrt{q}$ for $e$, $y^3 + 3\sqrt{q}y^2 + 2qy + r = 0$, every term of which is positive; therefore $\sqrt{q}$ is greater than the greatest value of $x$.

(303.) Cor. If the signs of the roots be changed, a limit greater than the greatest root of the resulting equation, with it's sign changed, is less than the least root of the proposed equation.

**Ex.** Required a limit less than the least root of the equation $y^3 - 3y + 72 = 0$.

When the signs of the roots are changed, this equation becomes $y^3 - 3y - 72 = 0$.

Assume $y = x + e$; then

$$\left. \begin{array}{r} x^3 + 3ex^2 + 3e^2x + e^3 \\ -\ 3x - 3e \\ -\ 72 \end{array} \right\} = 0,$$

and if 5 be substituted for $e$, every term becomes positive, consequently 5 is greater than the greatest root of the equation $y^3 - 3y - 72 = 0$; and $-5$ less than the least root of the equation $y^3 - 3y + 72 = 0$.

(304.) *The greatest negative coefficient increased by unity, is greater than the greatest root of an equation.*

Let $x^n - px^{n-1} - qx^{n-2} - \&c. = 0$, and if the coefficients be equal, $x^n - px^{n-1} - px^{n-2} - \&c. = 0$, or $x^n$
$$-p$$

$-p \times \overline{x^{n-1}+x^{n-2}\ldots\ldots+x+1} = 0$, that is, $x^n - p \times$

$\dfrac{x^n-1}{x-1} = 0$ (Art. 222). In this equation substitute

$\overline{1+p}$ for $x$, and the result is $\overline{1+p}|^n - p \times \dfrac{\overline{1+p}|^n - 1}{p}$,

or $+1$; and if any of the coefficients be positive, or less than $p$, the sum of the series, to be taken from $x^n$, will be diminished, and the result greater than before. Also, if for $x$, any quantity, still greater, be substituted, as $p + m + 1$, the result is $\overline{p+m+1}|^n - \dfrac{p}{p+m} \times$

$\overline{p+m+1}|^n + \dfrac{p}{p+m}$, or $\dfrac{m}{p+m} \times \overline{p+m+1}|^n + \dfrac{p}{p+m}$,

a positive quantity; therefore, $1 + p$ is greater than the greatest root (Art. 301).

(305.) In any equation, $x^n - px^{n-1} + qx^{n-2} - rx^{n-3} + sx^{n-4} - \&c. = 0$, whose roots are possible and positive,

$\sqrt{\dfrac{q}{n.\frac{n-1}{2}}}$; $\sqrt{\dfrac{p^2-2q}{n}}$; or $\sqrt[4]{\dfrac{q^2-2pr+2s}{n.\frac{n-1}{2}}}$ is less

than the greatest root.

Let $a$, $b$, $c$, &c. be the roots, taken in order; then $ab + ac + bc + \&c. = q$, and $a^2$ is greater than each of these products, therefore $n.\dfrac{n-1}{2} a^2$ (Art. 230), is

greater than $q$, or $a$ is greater than $\sqrt{n.\dfrac{q}{\frac{n-1}{2}}}$.

Also,

Also, $a^2 + b^2 + c^2 + \&c. = p^2 - 2q$ (Art. 297); and since $a$ is the greatest root, $na^2$ is greater than $a^2 + b^2 + c^2 + \&c.$ or $p^2 - 2q$; that is, $a$ is greater than

$$\sqrt{\frac{p^2 - 2q}{n}}.$$

Again, $q^2 - 2pr + 2s = a^2b^2 + a^2c^2 + b^2c^2 + \&c.$ (Art. 297), and $a^4$ is greater than each of these products, of which there are $n.\dfrac{n-1}{2}$; therefore $n.\dfrac{n-1}{2}a^4$ is greater than $a^2b^2 + a^2c^2 + b^2c^2 + \&c.$ or $q^2 - 2pr + 2s$; that is,

$a$ is greater than $\sqrt[4]{\dfrac{q^2 - 2pr + 2s}{n.\dfrac{n-1}{2}}}.$

(306.) Cor. If the equation have both positive and negative roots, and $-d$ be the least root; then, when $d$ is greater than the greatest root, it is greater than

$$\sqrt{\frac{p^2 - 2q}{n}}; \text{ or } \sqrt[4]{\frac{q^2 - 2pr + 2s}{n.\dfrac{n-1}{2}}}.$$

(307.) *The roots of the equation* $nx^{n-1} - \overline{n-1}.px^{n-2} + \overline{n-2}.qx^{n-3} - \&c. = 0$, *are limits between the roots of the equation* $x^n - px^{n-1} + qx^{n-2} - \&c. = 0$, *when the roots of the latter equation are possible.*

Let the roots of this equation, taken in order, be $a, b, c, -d, \&c.$ and in it, for $x$, substitute $y+e$, then by Art. 283,

$$\left.\begin{array}{l} y^n + ney^{n-1} \ldots\ldots + \underline{\quad} \quad ne^{n-1}y + e^n \\ \quad - py^{n-1} \ldots\ldots - \overline{n-1}.pe^{n-2}y - pe^{n-1} \\ \quad\quad\ldots\ldots + \overline{n-2}.qe^{n-3}y + qe^{n-2} \\ \quad\quad\quad - \&c. \end{array}\right\} = 0,$$

the

the roots of which equation are $a - e$, $b - e$, $c - e$, $-d - e$, &c. and the coefficient of the last term but one of any equation of $n$ dimensions, is the sum of the rectangles under $n - 1$ roots, with their signs changed (Art. 271); therefore,

$$\left. \begin{array}{l} ne^{n-1} - \overline{n-1}.pe^{n-2} + \overline{n-2}. \\ \quad q\,e^{n-3} - \&c. \end{array} \right\} = \left\{ \begin{array}{l} \overline{e - a}.\overline{e - b}.\overline{e - c}.\&c. \\ + \overline{e - a}.\overline{e - b}.\overline{e + d}.\&c. \\ + \overline{e - a}.\overline{e - c}.\overline{e + d}.\&c. \\ + \overline{e - b}.\overline{e - c}.\overline{e + d}.\&c. \end{array} \right.$$

in which, if $a$, $b$, $c$, $-d$, &c. be successively substituted for $e$, the results are,

$\overline{a - b}.\overline{a - c}.\overline{a + d}$. &c. which is positive,

$\overline{b - a}.\overline{b - c}.\overline{b + d}$. &c. negative,

$\overline{c - a}.\overline{c - b}.\overline{c + d}$. &c. positive,

$\overline{-d - a}.\overline{-d - b}.\overline{-d - c}$. &c. negative, &c.

therefore $a$, $b$, $c$, $-d$, are limits to the roots of the equation $ne^{n-1} - \overline{n-1}.pe^{n-2} + \&c. = 0$ (Art. 299), or, substituting $x$ for $e$, to the roots of the equation $nx^{n-1} - \overline{n-1}.px^{n-2} + \&c. = 0$. Let $\alpha$, $\beta$, $\gamma$, &c. be the roots of this equation, taken in order, then $a$, $\alpha$, $b$, $\beta$, $c$, $\gamma$, $-d$, &c. are arranged according to their magnitudes, that is, $\alpha$, $\beta$, $\gamma$, &c. lie between the roots of the equation $x^n - px^{n-1} + qx^{n-2} - \&c. = 0$.

(308.) Cor. 1. It appears from the preceding demonstration, that $nx^{n-1} - \overline{n-1}.px^{n-2} + \overline{n-2}.qx^{n-3} - \&c. = \overline{x - a}.\overline{x - b}.\overline{x - c}$. &c. $+ \overline{x - a}.\overline{x - b}.\overline{x + d}$. &c. $+ \overline{x - a}.\overline{x - c}.\overline{x + d}$. &c. $+ \overline{x - b}.\overline{x - c}.\overline{x + d}$. &c. where $a$, $b$, $c$, $-d$, &c. are the roots of the equation $x^n - px^{n-1} + qx^{n-2} - \&c. = 0$; also $x^n - px^{n-1} + qx^{n-2} - \&c. = \overline{x - a}.\overline{x - b}.\overline{x - c}.\overline{x + d}$. &c. therefore,

$$nx^{n-1}$$

$$\frac{nx^{n-1}-\overline{n-1}.px^{n-2}+\overline{n-2}.qx^{n-3}-\&c.}{x^n-px^{n-1}+qx^{n-2}-\&c.}=\frac{1}{x+d}+\frac{1}{x-c}$$

$$+\frac{1}{x-b}+\frac{1}{x-a}+\&c.$$

(309.) Cor. 2. If the limiting equation to $nx^{n-1}-$ $\overline{n-1}.px^{n-2}+\overline{n-2}.qx^{n-3}-\&c.=0$ be taken, it's roots will lie between the first and third, second and fourth, &c. roots of the first equation.

(310.) Cor. 3. The original equation has as many positive roots, and as many negative, as the limiting equation, and one more, which will be positive or negative according to the nature of the equation.

(311.) *Every equation whose roots are possible, has as many changes of signs from* $+$ *to* $-$, *and from* $-$ *to* $+$, *as it has positive roots; and as many continuations of the same sign, from* $+$ *to* $+$, *and from* $-$ *to* $-$, *as it has negative roots.*

Let $x^n-px^{n-1}\ldots\ldots\pm Sx^2\pm P.x\pm Q=0$, the equation of limits is $nx^{n-1}-\overline{n-1}.px^{n-2}\ldots\ldots\pm 2Sx$ $\pm P=0$, which, as far as it goes, has the same signs with the former; and therefore the original equation will have one more change of signs, or one more continuation of the same sign, than the limiting equation, according as the signs of $P$ and $Q$ are different, or the same.

Suppose $\alpha$, $\beta$, $\gamma$, &c. to be the roots of the limiting equation; then the roots of the original equation are, by Art. 310, of this form, $a$, $b$, $c$, $\pm d$, &c. therefore, $P$, with it's proper sign, $=n\times-a\times-\beta\times-\gamma\times$ &c. and $Q$ $=-a\times-b\times-c\times\mp d\times$ &c. (Art. 271), which products

ducts will have the same sign when the multiplier $d$ is positive, or the root $(-d)$ negative, and different signs when that root is positive. It appears then, that if the original equation have one more change of signs than the limiting equation, it has one more positive root; and, if it have one more continuation of the same sign, it has one more negative root; therefore if it can be shewn that every equation of $n-1$ dimensions, and consequently the equation $nx^{n-1} - \overline{n-1}.$ $px^{n-2} + \overline{n-2}.qx^{n-3} - \&c. = 0$, which is the limiting equation to $x^n - px^{n-1} + qx^{n-2} - \&c. = 0$, has as many changes of signs as it has positive roots, and as many continuations of the same sign as it has negative roots, the same rule will be true in the equation $x^n - px^{n-1} + qx^{n-2} - \&c. = 0$; or, in other words, if the rule be true of every equation of one order, it is true of every equation of the next superior order. Now in every simple equation $x - a = 0$, or $x + a = 0$, the rule is true, therefore it is true in every quadratic $x^2 \pm px \pm q = 0$; and if it be true in every quadratic, it is true in every cubic; and so on; that is, the rule is true in all cases.

In the demonstration, each root, $\pm d$, is supposed to be distinct from the rest, and a possible quantity.

Hence, when all the roots are possible, the number of positive roots is exactly known.

Ex. The equation $x^3 + x^2 - 14x + 8 = 0$, has two positive and one negative root; because the signs are $+, +, -, +$, in which there are two changes, one from $+$ to $-$, and the other from $-$ to $+$, and one continuation of the sign $+$.

(312.) *When any coefficient vanishes, it may be considered either as positive or negative, because the value of the whole expression is the same on either supposition.*

**Ex.** If the roots of the equation $x^3 - qx + r = 0$ be possible, two of them are positive and the third is negative; for there are two changes of signs in the equation $x^3 \pm o - qx + r = 0$, and one continuation of the same sign.

(313.) *If all the roots of an equation be positive, or all negative, and its terms be multiplied by the terms of any arithmetical progression, the resulting equation will be a limiting equation to the former.*

Let the roots of the equation $x^n - px^{n-1} + qx^{n-2} -$ &c. $= 0$, taken in order, be $a$, $b$, $c$, &c. and when they are substituted for $x$ in the quantity $nx^{n-1} - \overline{n-1}.px^{n-2} + $&c. let the results be $+L$, $-M$, $+N$, $-$ &c. (Art. 307.); then, when they are substituted in $Bx \times \overline{nx^{n-1} - n-1.px^{n-2}} + $ &c. the results will be $+BaL$, $-BbM$, $+BcN$, $-$ &c. but when the same quantities are substituted for $x$ in $x^n - px^{n-1} + qx^{n-2} -$ &c. or in $Ax^n - Apx^{n-1} + Aqx^{n-2} - $&c. the results are nothing; therefore when they are substituted in the sum of $Ax^n - Apx^{n-1} + Aqx^{n-2} - $&c. and $Bx \times \overline{nx^{n-1} - n-1.px^{n-2} + n-2.qx^{n-3}} - $&c. or in $\overline{A+nB}.$ $x^n - \overline{A+n-1}.B.px^{n-1} + \overline{A+n-2}.B.qx^{n-2} - $&c. the results are $+BaL$, $-BbM$, $+BcN$, $-$&c. therefore (Art. 299), $a$, $b$, $c$, &c. are limits of the roots of the equation $\overline{A+nB}.x^n - \overline{A+n-1}.B.px^{n-1} + \overline{A+n-2}.B.qx^{n-2} - $&c. $= 0$, which is deduced from the former by multiplying it's terms by the terms of the arithmetical progression $A+nB$, $A+\overline{n-1}. B$, $A+\overline{n-2}.B$, &c.

Conversely,

Conversely, the roots of this latter equation are limits to the roots of the former.

If $B$ be negative, or the series an increasing one, the results will be, $-BaL, +BbM, -BcN, +$&c. therefore $a, b, c,$ &c. are limits to the roots of the equation, $\overline{A-nB}.x^n - \overline{A-n-1}.B.px^{n-1} + \overline{A-n-2}.B.$ $qx^{n-2} -$ &c. $=0$, as before; only $a$ will be less than it's greatest root.

(314.) *If an equation have both positive and negative roots, and it's terms be multiplied by the terms of an arithmetical progression, an equation arises, whose roots are limits to the roots of the former, with this exception, that either two of it's roots, or none, lie between the positive and negative roots of the original equation, according as a decreasing or increasing progression is used.*

Let the roots of the equation $x^n - px^{n-1} - qx^{n-2} -$ &c. $=0$, taken in order, be $a, -b, -c,$ &c. When these values are substituted for $x$ in $nx^{n-1} - \overline{n-1}.px^{n-2}$ $-\overline{n-2}.qx^{n-3} -$ &c. the results are, $+L, -M, +N,$ $-$ &c. (Art. 307), therefore when they are substituted in $Bx \times \overline{nx^{n-1} - \overline{n-1}.px^{n-2} - \overline{n-2}.qx^{n-3}} -$ &c. or in $\overline{A+nB}.x^n - \overline{A+n-1}.B.px^{n-1} - \overline{A+n-2}.B.qx^{n-2}$ $-$ &c. the results are $+BaL, +BbM, -BcN, +$&c. Now the roots of the equation $\overline{A+nB}.x^n - \overline{A+n-1}.B.$ $px^{n-1} -$ &c. $=0$ are of the same form with the roots of the original equation*, because there is the same number of changes of signs in both ; let these roots, taken in order, be $a, -\beta, -\gamma,$ &c. and since both

$a,$ and

---

* Here we suppose that the first term is not taken out, and that the signs of the terms of the progression are not changed.

$a$, and $-b$, when substituted in the limit, give positive results, either two roots $a$, and $-\beta$, or none, lie between them (Art. 300); and $-b$, a negative quantity, cannot be greater than $a$, a positive one; therefore the order of the magnitudes is $a$, $a$, $-\beta$, $-b$, $-\gamma$, $-c$, &c. that is, when the terms of the equation are multiplied by the terms of the decreasing arithmetical progression, $A + nB$, $A + \overline{n-1}.B$, $A + \overline{n-2}.B$, &c. two roots, $a$, and $-\beta$, of the limiting equation, lie between the positive and negative roots, $a$ and $-b$, of the original equation.

When $B$ is negative, and $nB$ less than $A$, or the series an ascending one, it may be proved, as before, that when $a$, $-b$, $-c$, &c. are substituted for $x$, in $\overline{A-nB}.x^n - \overline{A-n-1}.B.px^{n-1} - \overline{A-n-2}.B.qx^{n-2} -$ &c. the results are $-BaL$, $-BbM$, $+BcN$, $-$ &c. therefore $a$ is less than $\alpha$, and either two roots $-\beta$, $-\gamma$, or none, lie between $a$ and $-b$; there cannot be two, because then $-b$, and $-c$, are both less than $-\gamma$, and when substituted for $x$, must give results affected with the same sign; but the results are $-BbM$, and $+BcN$; therefore the order of the roots is $a$, $a$, $-b$, $-\beta$, $-c$, $-\gamma$, $-$ &c. that is, no root of the limiting equation lies between $a$, and $-b$. $\big($ If $A$ be less than $nB$, the first term of the limiting equation is negative, and the signs of all the terms being changed, to reduce the equation to a proper form (Art. 262), the series becomes a decreasing one. $\big)$

(315.) In the preceding demonstration, the limiting equation is supposed to contain as many roots as the original equation, and of the same form. If the

arithmetical

arithmetical progression begin from nothing, it may be
proved in the same manner, that no root of the limit,
thus deduced, lies between the positive and negative
roots of the proposed equation.

(316.) Ex. Let the proposed equation be $x^3 - qx + r$
$= 0$. The roots of the equation $3x^2 - q = 0$, are
limits which lie between it's roots (Art. 307.)

Let the terms of the equation be multiplied by
the series, 3, 2, 1, 0, and the limiting equation is

$3x^3 - qx = 0$, whose roots are $\sqrt{\dfrac{q}{3}}$, 0, $-\sqrt{\dfrac{q}{3}}$, two

of which, 0 and $-\sqrt{\dfrac{q}{3}}$ lie between the positive and

negative roots of the proposed equation (Art. 314).

Let the terms be multiplied by the series 0, $-1$,
$-2$, $-3$, the limiting equation thus obtained is
$2qx - 3r = 0$, whose root $\dfrac{3r}{2q}$ lies between the positive
roots of the equation $x^3 - qx + r = 0$ (Art. 315).

(317.) *To find between which of the roots, of a
proposed equation, any given number lies.*

Let the roots of the proposed equation be diminished by the given number, and the number of
negative roots, in the transformed equation, will shew
it's place among the roots of the original equation.

Ex. To find between which of the roots of the
equation $x^3 - 9x^2 + 23x - 15 = 0$, the number 2 lies.

Assume

Assume $x = y + 2$; then,

$$\left.\begin{array}{r} x^3 \\ -9x^2 \\ +23x \\ -15 \end{array}\right\} = \left.\begin{array}{r} y^3 + 6y^2 + 12y + 8 \\ -9y^2 - 36y - 36 \\ +23y + 46 \\ -15 \end{array}\right\} = 0,$$

or $y^3 - 3y^2 - y + 3 = 0$, which has one negative root; and the roots of the proposed equation are all positive; therefore two of them are greater, and one is less, than 2.

(318.) In general, the last term, and the coefficients of the other terms of the transformed equation are found by substituting the number, by which the roots are to be diminished, for $x$, in the quantities,

$$x^n - p x^{n-1} + q x^{n-2} - \&c.$$
$$n x^{n-1} - \overline{n-1} . p x^{n-2} + \overline{n-2} . q x^{n-3} - \&c.$$
$$n . \frac{n-1}{2} x^{n-2} - \overline{n-1} . \frac{n-2}{2} p x^{n-3} + \overline{n-2} . \frac{n-3}{2} q x^{n-4} - \&c.$$
$$\&c.$$

See Art. 283. And by substituting, successively, different numbers for $x$, the limits of the roots of the proposed equation may be found.

Ex. Let the proposed equation be $x^4 - 2x^3 - 5x^2 + 10x - 3 = 0.$

| values of $x$ | 1 | 2 | 3 |
|---|---|---|---|
| $x^4 - 2x^3 - 5x^2 + 10x - 3 =$ | $+1$ | $-3$ | $+9$ |
| $4x^3 - 6x^2 - 10x + 10. \ldots =$ | $-2$ | $-2$ | $+34$ |
| $6x^2 - 6x - 5 \ldots\ldots\ldots =$ | $-5$ | $+7$ | $+31$ |
| $4x - 2 \ldots\ldots\ldots\ldots =$ | $+2$ | $+6$ | $+10$ |
| $1 \ldots\ldots\ldots\ldots\ldots =$ | $+1$ | $+1$ | $+1$ |

From

From the changes of signs in the proposed equation, it appears, that it has one negative and three positive roots; when these roots are diminished by 1, they become two positive and two negative; when diminished by 2, they become one positive and three negative; and when diminished by 3, they all become negative; therefore one root of the proposed equation lies between 0 and 1, one between 1 and 2, and the third between 2 and 3.

By changing the signs of the roots, and proceeding in the same way, we may find, that the negative root lies between $-2$, and $-3$.

## ON THE DEPRESSION AND SOLUTION
## OF EQUATIONS.

(319.) *If an equation contain equal roots, these may be found, and the equation reduced as many dimensions lower as there are equal roots.*

Let the roots of the equation $x^n - px^{n-1} + qx^{n-2} -$ &c. $= 0$, be $a$, $b$, $c$, $d$, &c. then (Art. 307),

$$\left. \begin{array}{l} nx^{n-1} - \overline{n-1}.px^{n-2} + \\ \overline{n-2}.qx^{n-3} - \&c. \end{array} \right\} = \left\{ \begin{array}{l} \overline{x-a}.\overline{x-b}.\overline{x-c}. \&c. \\ + \overline{x-a}.\overline{x-b}.\overline{x-d}. \&c. \\ + \overline{x-a}.\overline{x-c}.\overline{x-d}. \&c. \\ + \overline{x-b}.\overline{x-c}.\overline{x-d}. \&c. \end{array} \right.$$

Suppose $a = b$; then,

$$nx^{n-1}$$

$$n x^{n-1} - \overline{n-1}.p x^{n-2} + \atop n-2.q x^{n-3} - \&c. \Big\} = \begin{cases} \overline{x-a}.\overline{x-a}.\overline{x-c}. \ \&c. \\ + \overline{x-a}.\overline{x-a}.\overline{x-d}. \ \&c. \\ + \overline{x-a}.\overline{x-c}.\overline{x-d}. \ \&c. \\ + \overline{x-a}.\overline{x-c}.\overline{x-d}. \ \&c. \end{cases}$$

the whole of which is divisible by $x-a$ without remainder, that is, $a$ is a root of the equation $n x^{n-1} - \overline{n-1}.p x^{n-2} + \overline{n-2}.q x^{n-3} - \&c. = 0.$*

If three roots $a$, $b$, $c$, be equal, $\overline{x-a}.\overline{x-a}$ will be found in every product, therefore the equation is divisible by $\overline{x-a}.\overline{x-a}$ without remainder, or two of it's roots are $a$, $a$ (Art. 268).

In the same manner, if the original equation have $m$ equal roots, the equation $n x^{n-1} - \overline{n-1}.p x^{n-2} + \overline{n-2}.q x^{n-3} - \&c. = 0$, has $m-1$ of those roots.

(320.) Hence it appears, that when there are $m$ equal roots, the two equations have a common measure of the form $\overline{x-a}^{m-1}$, which may be obtained in the usual way (Art. 90), and $m$ roots of the original equation may thus be known. Divide this equation by $\overline{x-a}^{m}$, and the resulting equation, of $n-m$ dimensions, contains the other roots.

Ex. Let the equation $x^3 - p x^2 + q x - r = 0$, have two equal roots; then $3 x^2 - 2 p x + q = 0$ has one of them; and the two equations have a common measure which is a simple equation; consequently, the quantities $9 x^3 - 9 p x^2 + 9 q x - 9 r$ and $3 x^2 - 2 p x + q$, have the same common measure, which is thus found,

$$3 x^2 -$$

---

* The equation $\overline{A + n B}.x^n - \overline{A + n - 1}.B.p x^{n-1} + \&c = 0$ has also one of the equal roots. See Art. 313.

$$3x^2 - 2px + q)9x^3 - 9px^2 + 9qx - 9r(3x - p$$
$$9x^3 - 6px^2 + 3qx$$

$$\overline{\phantom{9x^3 - 6px^2 + 3qx}}$$
$$-3px^2 + 6qx - 9r$$
$$-3px^2 + 2p^2x - pq$$

$$\overline{6q - 2p^2} \times x + pq - 9r$$

hence, $\overline{6q - 2p^2}.x + pq - 9r$ is a divisor of the equation $x^3 - px^2 + qx - r = 0$; that is, $\overline{6q - 2p^2}.x + pq - 9r = 0$, and $x = \dfrac{9r - pq}{6q - 2p^2}$ (Art. 268).

Thus two roots of the equation are discovered ; and since $p$ is the sum of all the roots, the third root is the difference between $p$ and the sum of the two equal roots.

Let the proposed equation be $x^3 - 4x^2 + 5x - 2 = 0$.

Here, $p = 4$, $q = 5$, $r = 2$; and one of the equal roots is $\dfrac{9r - pq}{6q - 2p^2} = 1$; and the third root is $p - 2 = 4 - 2 = 2$.

But it must be observed, that though $\dfrac{9r - pq}{6q - 2p^2}$ should be found, upon trial, to be a root of the proposed cubic, this equation has not two equal roots, unless $\dfrac{9r - pq}{6q - 2p^2}$ be a root of the depressed equation $3x^2 - 2px + q = 0$.

(321.) If the roots of the equation $x^3 - px^2 + qx - r = 0$, be in arithmetical progression, $\dfrac{9r - pq}{6q - 2p^2}$ is one of them.

Let

Let the roots be $a - b$, $a$, $a + b$; then $p = 3a$, $q = 3a^2 - b^2$, and $r = a^3 - ab^2$; hence, $9r - pq = -6ab^2$, and $6q - 2p^2 = -6b^2$; therefore, $\dfrac{9r - pq}{6q - 2p^2} = \dfrac{-6ab^2}{-6b^2} = a$.

(322.) *If two roots of an equation be of the form* + a, − a, *differing only in their signs, they may be found, and the equation depressed.*

Change the signs of the roots, and the resulting equation has two roots $+a$, $-a$; thus we have two equations with a common measure, $x^2 - a^2$, which may be found, and the equation depressed, as in the preceding case.

Ex. Required the roots of the equation $x^4 + 3x^3 - 7x^2 - 27x - 18 = 0$, two of which are of the form $+a$, $-a$. By changing the signs of the alternate terms we obtain the equation $x^4 - 3x^3 - 7x^2 + 27x - 18 = 0$, which has two roots of the form $+a$, $-a$; and the common quadratic divisor, of the two equations, is $x^2 - 9 = 0$, hence, $x = \pm 3$. To obtain the other roots, divide $x^4 + 3x^3 - 7x^2 - 27x - 18 = 0$, by $x^2 - 9 = 0$, and the roots of the resulting equation $x^2 + 3x + 2 = 0$, are the roots sought.

(323.) By this method, when the coefficients are rational, surd roots of the form $\pm \sqrt{a}$ may be discovered (See Art. 279).

(324.) When there are two other roots of the same form, the equations will have a common divisor $x^4 - Qx^2 + R = 0$, which contains the four roots, $a$, $-a$, $b$, $-b$.

It

If the roots of an equation have any other given relation, they may be found in a similar manner (See Waring's *Alg.* Cap. 3); but as particular relations of the roots to each other are rarely known, it seems unnecessary to prosecute the subject farther, in this place.

## SOLUTION OF RECURRING EQUATIONS.

(325.) *The roots of a recurring equation, of an even number of dimensions, exceeding a quadratic, may be found by the solution of an equation of half the number of dimensions.*

Let $x^n - px^{n-1} \ldots \ldots - px + 1 = 0$; its roots are of the form $a, \frac{1}{a}, b, \frac{1}{b}$, &c. (Art. 292); or it may be conceived to be made up of quadratic factors, $\overline{x - a}.$ $\overline{x - \frac{1}{a}}; \overline{x - b}.\overline{x - \frac{1}{b}}$; &c. *i. e.* if $m = a + \frac{1}{a}, n = b + \frac{1}{b}$, &c. of the quadratic factors, $x^2 - mx + 1$; $x^2 - nx + 1$; &c. Then by multiplying these together, and equating the coefficients with those of the proposed equation, the values of $m$, $n$, &c. may be found. Moreover, for every value of $m$ there are two values of $x$; therefore the equation for determining the value of $m$, will rise only to half as many dimensions as $x$ rises in the original equation.

(326.) If

(326.) If the recurring equation be of an *odd* number of dimensions, $+1$, or $-1$ is a root (Art. 295); and the equation may therefore be reduced to one of the same kind, of an even number of dimensions, by division.

Ex. 1. Let $x^3 - 1 = 0$. Unity is one root of this equation, and by dividing $x^3 - 1$ by $x - 1$, the equation $x^2 + x + 1 = 0$ is obtained, which contains the other two roots, $\dfrac{-1+\sqrt{-3}}{2}$, and $\dfrac{-1-\sqrt{-3}}{2}$

$$x^2 + x + 1 = 0$$

$$x^2 + x + \frac{1}{4} = \frac{1}{4} - 1 = \frac{-3}{4}$$

$$x + \frac{1}{2} = \pm \frac{\sqrt{-3}}{2},$$

or $x = \dfrac{-1 \pm \sqrt{-3}}{2}$; that is, the three roots of the equation $x^3 - 1 = 0$, or the three cube roots of 1, are 1, $\dfrac{-1+\sqrt{-3}}{2}$, and $\dfrac{-1-\sqrt{-3}}{2}$.

In the same manner, the roots of the equation $x^3 + 1 = 0$ are found to be $-1$, $\dfrac{1+\sqrt{-3}}{2}$, and $\dfrac{1-\sqrt{-3}}{2}$. This also follows from Art. 281.

Ex. 2. Let $x^4 - 1 = 0$. Two roots of this equation are $+1$, $-1$; and by division, $\dfrac{x^4 - 1}{x^2 - 1} = x^2 + 1 = 0$, an equation which contains the other two roots, $+\sqrt{-1}$, and $-\sqrt{-1}$.

Ex. 3.

Ex. 3. Let $x^4 + 1 = 0$. Assume $\overline{x^2 - mx + 1} \times \overline{x^2 - nx + 1} = x^4 + 1$; that is, $x^4 - \overline{m+n}.x^3 + \overline{mn+2}.x^2 - \overline{m+n}.x + 1 = x^4 + 1$, and by equating the coefficients, $m + n = 0$, and $mn + 2 = 0$; hence $n = -m$, and $-m^2 + 2 = 0$, or $m^2 = 2$, and $m = \pm\sqrt{2}$. Therefore the two quadratics which contain the roots of the biquadratic, are $x^2 + \sqrt{2}.x + 1 = 0$, and $x^2 - \sqrt{2}.x + 1 = 0$, from the solution of which it appears that the roots are

$$\frac{-1 \pm \sqrt{-1}}{\sqrt{2}} \text{ and } \frac{1 \pm \sqrt{-1}}{\sqrt{2}}.$$

In the same manner may the roots of the equations $x^5 + 1 = 0$, and $x^6 + 1 = 0$, be found.

# THE SOLUTION OF A CUBIC EQUATION BY CARDAN'S RULE.

(327.) Let the equation be reduced to the form $x^3 - qx + r = 0$, where $q$ and $r$ may be positive or negative.

Assume $x = a + b$, then the equation becomes $\overline{a+b}|^3 - q \times \overline{a+b} + r = 0$, or $a^3 + b^3 + 3ab \times \overline{a+b} - q \times \overline{a+b} + r = 0$; and since we have two unknown quantities, $a$ and $b$, and have made only one supposition respecting them, viz. that $a + b = x$, we are at liberty to make another; let $3ab - q = 0$, then the equation becomes $a^3 + b^3 + r = 0$; also, since $3ab - q = 0$, $b = \dfrac{q}{3a}$, and by

substitution,

substitution, $a^3 + \dfrac{q^3}{27\,a^3} + r = 0$, or $a^6 + r\,a^3 + \dfrac{q^3}{27} = 0$, an equation of a quadratic form; and by completing the square, $a^6 + r\,a^3 + \dfrac{r^2}{4} = \dfrac{r^2}{4} - \dfrac{q^3}{27}$, and $a^3 + \dfrac{r}{2} = \pm$ $\sqrt{\dfrac{r^2}{4} - \dfrac{q^3}{27}}$; therefore $a^3 = -\dfrac{r}{2} \pm \sqrt{\dfrac{r^2}{4} - \dfrac{q^3}{27}}$, and $a = \sqrt[3]{-\dfrac{r}{2} \pm \sqrt{\dfrac{r^2}{4} - \dfrac{q^3}{27}}}$. Also, since $a^3 + b^3 + r = 0$, $b^3 = -\dfrac{r}{2} \mp \sqrt{\dfrac{r^2}{4} - \dfrac{q^3}{27}}$, and $b = \sqrt[3]{-\dfrac{r}{2} \mp \sqrt{\dfrac{r^2}{4} - \dfrac{q^3}{27}}}$; therefore $x = a + b = \sqrt[3]{-\dfrac{r}{2} \pm \sqrt{\dfrac{r^2}{4} - \dfrac{q^3}{27}}} + \sqrt[3]{-\dfrac{r}{2} \mp \sqrt{\dfrac{r^2}{4} - \dfrac{q^3}{27}}}$.

We may observe, that when the sign of $\sqrt{\dfrac{r^2}{4} - \dfrac{q^3}{27}}$, in one part of the expression, is positive, it is negative in the other, that is $x = \sqrt[3]{-\dfrac{r}{2} + \sqrt{\dfrac{r^2}{4} - \dfrac{q^3}{27}}} + \sqrt[3]{-\dfrac{r}{2} - \sqrt{\dfrac{r^2}{4} - \dfrac{q^3}{27}}}$.

(328.) Since $b = \dfrac{q}{3\,a}$, the value of $x$ is also

$$\sqrt[3]{-\dfrac{r}{2} \pm \sqrt{\dfrac{r^2}{4} - \dfrac{q^3}{27}}} + \dfrac{q}{3\sqrt[3]{-\dfrac{r}{2} \pm \sqrt{\dfrac{r^2}{4} - \dfrac{q^3}{27}}}}$$

Ex.

172      CARDAN'S RULE.

## Ex.

Let $x^3 + 6x - 20 = 0$; here $q = -6$, $r = -20$,

$$x = \sqrt[3]{10 + \sqrt{108}} + \sqrt[3]{10 - \sqrt{108}} = 2.732 - .732$$
$$= 2\,*.$$

(329.) Cor. 1. Having obtained one value of $x$, the equation may be depressed to a quadratic, and the other roots found (Art. 267).

(330.) Cor. 2. The possible values of $a$ and $b$ being discovered, the other roots are known without the solution of a quadratic.

The values of the cube roots of $a^3$ are $a$, $\dfrac{-1 + \sqrt{-3}}{2} a$, and $\dfrac{-1 - \sqrt{-3}}{2} a$ ; and the values of

the cube root of $b^3$, are $b$, $\dfrac{-1 + \sqrt{-3}}{2} b$, $\dfrac{-1 - \sqrt{-3}}{2} b$

(Art. 326). Hence it appears, that there are nine values of $a + b$, three only of which can answer the conditions of the equation, the others having been introduced by involution. These nine values are,

1.   $a + b$

2.   $a + \dfrac{-1 + \sqrt{-3}}{2} b$

3.   $a + \dfrac{-1 - \sqrt{-3}}{2} b$

4.   $\dfrac{-1 + \sqrt{-3}}{2} a + b$

5.   $-1$

---

* See Art. 259.

5. $\dfrac{-1+\sqrt{-3}}{2}a+\dfrac{-1+\sqrt{-3}}{2}b$

6. $\dfrac{-1+\sqrt{-3}}{2}a+\dfrac{-1-\sqrt{-3}}{2}b$

7. $\dfrac{-1-\sqrt{-3}}{2}a+b$

8. $\dfrac{-1-\sqrt{-3}}{2}a+\dfrac{-1+\sqrt{-3}}{2}b$

9. $\dfrac{-1-\sqrt{-3}}{2}a+\dfrac{-1-\sqrt{-3}}{2}b.$

In the operation, we assume $3ab=q$, that is, the product of the corresponding values of $a$ and $b$ is supposed to be possible. This consideration excludes the 2d. 3d. 4th. 5th. 7th. and 9th. values of $a+b$, or $x$; therefore the three roots of the equation are $a+b$,

$$\dfrac{-1+\sqrt{-3}}{2}a+\dfrac{-1-\sqrt{-3}}{2}b, \text{ and } \dfrac{-1-\sqrt{-3}}{2}a+$$

$$\dfrac{-1+\sqrt{-3}}{2}b.$$

The value of $x$ is also $a+\dfrac{q}{3a}$; therefore, if $a$, $\alpha a$, $\beta a$ be the three roots of $a^3$, the roots of the cubic are

$$a+\dfrac{q}{3a}; \; \alpha a+\dfrac{q}{3\alpha a}; \; \beta a+\dfrac{q}{3\beta a}.$$

(331.) Cor. 3. This solution only extends to those cases in which the cubic has two impossible roots.

For if the roots be $m+\sqrt{3n}$, $m-\sqrt{3n}$, and $-2m$, then $-q$ (the sum of the products of every two with their signs changed) $=-3m^2-3n$, and $\dfrac{q}{3}=m^2+n$;

also,

also, $r$ (the product of all the roots with their signs changed) $=2m^3-6mn$, and $\dfrac{r}{2}=m^3-3mn$; and by involution,

$$\frac{r^2}{4}=m^6-6m^4n+9m^2n^2$$

$$\frac{q^3}{27}=m^6+3m^4n+3m^2n^2+n^3$$

Hence, $\dfrac{r^2}{4}-\dfrac{q^3}{27}=-9m^4n+6m^2n^2-n^3=$

$-n\times\overline{9m^4-6m^2n+n^2}$, and $\sqrt{\dfrac{r^2}{4}-\dfrac{q^3}{27}}=\sqrt{-n\times\overline{3m^2-n}}$,

a quantity manifestly impossible, unless $n$ be negative, that is, unless two roots of the proposed cubic be impossible.

## SOLUTION OF A BIQUADRATIC BY DES CARTES's METHOD.

(332.) Any biquadratic may be reduced to the form $x^4+qx^2+rx+s=0$, by taking away the second term (Art. 284). Suppose this to be made up of the two quadratics, $x^2+ex+f=0$, and $x^2-ex+g=0$, where $+e$ and $-e$ are made the coefficients of the second terms, because the second term of the biquadratic is wanting, that is, the sum of it's roots is 0. By multiplying these quadratics together, we have $x^4+\overline{g+f-e^2}$. $x^2+\overline{eg-ef}.x+fg=0$, which equation is made to coincide with the former, by equating their coefficients, or making $g+f-e^2=q$, $eg-ef=r$, and $fg=s$; hence, $g+f=q+e^2$, also $g-f=\dfrac{r}{e}$, and by taking the sum and difference of these equations, $2g=q+e^2+\dfrac{r}{e}$, and

$\dfrac{r}{e}$, and $2f = q + e^2 - \dfrac{r}{e}$; therefore $4fg = q^2 + 2qe^2 + e^4 -$

$\dfrac{r^2}{e^2} = 4s$, and multiplying by $e^2$, and arranging the terms according to the dimensions of $e$, $e^6 + 2qe^4 + \overline{q^2 - 4s} \times e^2 - r^2 = 0$ ; or, making $y = e^2$, $y^3 + 2qy^2 + \overline{q^2 - 4s}.y - r^2 = 0$.

By the solution of this cubic, a value of $y$, and therefore of $\sqrt{y}$, or $e$, is obtained; also $f$ and $g$, which are respectively equal to $\dfrac{q + e^2 - \dfrac{r}{e}}{2}$ and $\dfrac{q + e^2 + \dfrac{r}{e}}{2}$, are known; the biquadratic is thus resolved into two quadratics, whose roots may be found.

It may be observed, that which ever value of $y$ is used, the same values of $x$ are obtained.

(333.) This solution can only be applied to those cases, in which two roots of the biquadratic are possible and two impossible.

Let the roots be $a$, $b$, $c$, $-\overline{a + b + c}$; then since $e$, the coefficient of the second term of one of the reducing quadratics, is the sum of two roots, it's different values are $a + b$, $a + c$, $b + c$, $-\overline{a+b}$, $-\overline{a+c}$, $-\overline{b+c}$, and the values of $e^2$, or $y$, are $\overline{a + b}|^2$, $\overline{a + c}|^2$, $\overline{b + c}|$ ; all of which being possible, the cubic cannot be solved by any direct method. Suppose the roots of the biquadratic to be $a + b\sqrt{-1}, a - b\sqrt{-1}, -a + c\sqrt{-1}$, $-a - c\sqrt{-1}$; the values of $e$ are $2a$, $\overline{b + c}.\sqrt{-1}$, $\overline{b - c}.\sqrt{-1}$, $-\overline{b - c}.\sqrt{-1}$, $-\overline{b + c}.\sqrt{-1}$ and $-2a$; and the three values of $y$ are, $\overline{2a}|^2$, $-\overline{b + c}|^2$, $-\overline{b - c}|^2$, which

which are all possible, as in the preceding case. But if the roots of the biquadratic be $a + b\sqrt{-1}$, $a - b\sqrt{-1}$, $-a+c$, $-a-c$, the values of $y$ are $\overline{2a}|^2$, $\overline{c+b\sqrt{-1}}|^2$, $\overline{c-b\sqrt{-1}}|^2$, two of which are impossible; therefore the cubic may be solved by Cardan's rule.

## Dr. WARING's SOLUTION.

(334.) Let the proposed biquadratic be $x^4 + 2px^3 = qx^2 + rx + s$; now $\overline{x^2+px+n}|^2 = x^4 + 2px^3 + \overline{p^2 + 2n}.x^2 + 2pnx + n^2$, if therefore $\overline{p^2 + 2n}.x^2 + 2pnx + n^2$ be added to both sides of the proposed biquadratic, the first part is a complete square, $\overline{x^2+px+n}|^2$, and the latter part, $\overline{p^2 + 2n + q} \times x^2 + \overline{2pn+r} \times x + n^2 + s$, is a complete square, if $4 \times \overline{p^2 + 2n + q} \times \overline{n^2 + s} = \overline{2pn+r}|^2$ (Art. 127), that is, multiplying and arranging the terms according to the dimensions of $n$, if $8n^3 + 4qn^2 + \overline{8s - 4rp}.n + 4qs + 4p^2s - r^2 = 0$. From this equation, let a value of $n$ be obtained and substituted in the equation $\overline{x^2 + px + n}|^2 = \overline{p^2 + 2n + q}.x^2 + \overline{2pn + r}.x + n^2 + s$; then extracting the square root on both sides, $x^2 + px + n = \pm \sqrt{p^2 + 2n + q} \times x + \sqrt{n^2 + s}$, when $2pn + r$ is positive; or $x^2 + px + n = \pm \sqrt{p^2 + 2n + q}.x - \sqrt{n^2 + s}$, when $2pn + r$ is negative; and from these two quadratics, the four roots of the given biquadratic may be determined.

Ex.

Ex. Let $x^4 - 6x^3 + 5x^2 + 2x - 10 = 0$ be the proposed equation.

By comparing this with the equation $x^4 + 2px^3 - qx^2 - rx - s = 0$, we have $2p = -6$, or $p = -3$, $q = -5$, $r = -2$, $s = 10$; and $8n^3 + 4qn^2 + \overline{8s - 4rp} \times n + 4qs + 4p^2s - r^2 = 0$, is $8n^3 - 20n^2 + 56n + 156 = 0$, or $2n^3 - 5n^2 + 14n + 39 = 0$, one of whose roots is $-\dfrac{3}{2}$;

hence, $\overline{x^2 - 3x - \dfrac{3}{2}}\Big|^2 = x^2 + 7x + \dfrac{49}{4}$, and $x^2 - 3x - \dfrac{3}{2} = \pm\overline{x + \dfrac{7}{2}}$; or $x^2 - 4x - 5 = 0$, and $x^2 - 2x + 2 = 0$; the roots of these quadratics, $-1$, $5$, $1 + \sqrt{-1}$, $1 - \sqrt{-1}$, are the roots of the proposed biquadratic.

(335.) This solution can only be applied to those cases in which two roots of the biquadratic are possible, and two impossible.

Let the roots be $a$, $b$, $c$, $d$, then $n - \sqrt{n^2 + s}$, the last term of one of the quadratics, to which the equation is reduced, is the product of two of them, as $ab$; therefore $n - ab = \sqrt{n^2 + s}$, and squaring both sides, $n^2 - 2nab + a^2b^2 = n^2 + s$, or $-2nab + a^2b^2 = s = -abcd$ (Art. 271), and dividing both sides by $-ab$, $2n - ab = cd$, or $2n = ab + cd$, and $n = \dfrac{ab + cd}{2}$; the other values of $n$ are $\dfrac{ac + bd}{2}$, and $\dfrac{ad + bc}{2}$, therefore, when $a$, $b$, $c$, $d$, are possible, the values of $n$ are possible. Also, when these quantities are all impossible, the values of $n$ are all possible; in neither case therefore

M      can

can the value of $n$ be obtained by Cardan's rule; but if two roots of the biquadratic be possible and two roots impossible, two values of $n$ will be impossible, and the cubic may be solved, and consequently the roots of the proposed equation may be found.

## THE METHOD OF DIVISORS.

(336.) *Since the last term of an equation is the product of all the roots with their signs changed, if any root be a whole number, it may be found amongst the divisors of the last term.*

Ex. Suppose $x^3 - 4x^2 - 6x + 12 = 0$; the divisors of the last term are 1, $-1$, 2, $-2$, 3, $-3$, 4, $-4$, 6, $-6$, 12, $-12$, and by substituting these successively for $x$, we find that $-2$ is a root of the equation.

(337.) When the last term admits of many divisors, the number of trials may be lessened by finding the limits between which the roots of the equation lie; or by increasing, or diminishing, the roots of the equation, and thus lessening the number of divisors of the last term.

(338.) *The number of trials may also be lessened by substituting three or more terms of the arithmetical progression 1, 0, $-1$, &c. for the unknown quantity, and forming the divisors of the results, taken in order, into arithmetical progressions, in which the common difference is unity ; as it will only be necessary to try those divisors of the last term of the equation which are found in these progressions.*

Let

Let $\overline{x+a}.\mathbf{Q}=0$ be the equation, one factor of which is $x+a$, and $\mathbf{Q}$ the product of the rest; if 1, 0, $-1$, be successively substituted for $x$, the results are respectively divisible by $a+1$, $a$, and $a-1$; therefore amongst the divisors of the results, formed into arithmetical progressions in which the common difference is unity, is found the decreasing progression $a+1$, $a$, $a-1$; and if all the terms corresponding to $a$, with their signs changed, be substituted in the equation for $x$, the integral values of $x$ will be discovered.

Ex. Let the proposed equation be $x^3 - 4x^2 - 6x + 12 = 0$.

| Supp. | Results. | Divisors. | Progr. |
|---|---|---|---|
| $x=\ \ \ 1$ | 3 | 1, 3 | 3 |
| $x=\ \ \ 0$ | 12 | 1, 2, 3, 4, 6, 12 | 2 |
| $x=-1$ | 13 | 1, 13 | 1 |

The only decreasing progression that can be formed out of the divisors is 3, 2, 1, therefore if one root of the equation be a whole number, $-2$ is that root; and on trial it is found to succeed.

(339.) If the highest power of the unknown quantity be affected with a coefficient, let $\overline{mx+a} \times \mathbf{Q}=0$ be the equation, and substitute 1, 0, $-1$, successively for $x$, then $a+m$, $a$, and $a-m$ are divisors of the results, if the equation have a factor of the form $mx+a$, or a root $-\dfrac{a}{m}$. Also $m$, the common difference, in the arithmetical progression $a+m$, $a$, $a-m$, is a divisor of the coefficient of the first term of the

M 2                     equation.

equation. In this case therefore, all the decreasing progressions must be taken out of the divisors of the resulting quantities, in which the common difference is unity, or some divisor of the coefficient of the highest term of the equation, and amongst them is the progression $a+m$, $a$, $a-m$; therefore by making trial, successively, of the terms corresponding to $a$ in the progressions thus obtained, the factor $mx+a$, which divides the equation without remainder, will be found.

Ex. To find a divisor of the equation $8x^3 - 26x^2 + 11x + 10 = 0$, if it admit one of the form $mx + a$.

| Sup. | Res | Divisors. | Progress. |
|---|---|---|---|
| $x = 1$ | 3 | $1, 3, -1, -3$ | $3, 3, -3$ |
| $x = 0$ | 10 | $1, 2, 5, 10, -1, -2, -5, -10$ | $1, 2, -5$ |
| $x = -1$ | $-35$ | $1, 5, 7, 35, -1, -5, -7, -35$ | $-1, 1, -7$ |

The decreasing progressions, in which the common difference is a divisor of 8, formed out of the divisors, are $3, 1, -1$; $3, 2, 1$; and $-3, -5, -7$; therefore the factors to be tried are $2x+1$, $x+2$, and $2x-5$, the last of which succeeds, and consequently $2x-5$,

$= 0$ (Art. 268), or $x = \dfrac{5}{2}$.

(340.) If an equation be of four, or more dimensions, though it has no divisor of the form $mx + a$, it may have one of the form $\pm mx^2 \pm nx \pm r$.

To find when this is the case, let $\pm \overline{mx^2 + nx + r}$ $\times Q = 0$ be the equation; and for $x$ substitute successively, $p + e$, $p$, $p - e$, &c. then $\pm m.\overline{p + e}|^2 + n.\overline{p + e} + r$, $\pm mp^2 + np + r$, $\pm m.\overline{p - e}|^2 + n.\overline{p - e} + r$, &c. are divisors

sors of the resulting quantities; and if they be respectively subtracted from, or added to, $m.\overline{p+e}|^2$, $m.p^2$, $m.\overline{p-e}|^2$, &c. the remainders, or sums, are $n.\overline{p+e}+r$, $np+r$, $n.\overline{p-e}+r$, &c. which form a decreasing arithmetical progression whose common difference is $ne$. When $p=0$, and $e=1$, this progression becomes $n+r$, $r$, $-n+r$ &c. and in all cases $m$ is a divisor of the first term of the equation. Let therefore $1$, $0$, $-1$, $-2$, &c. be substituted for $x$ in the proposed equation, and let the differences and sums of the divisors of the results, and $m$, $0$, $m$, $4m$, &c. be taken; then if all the arithmetical progressions possible be formed out of these quantities, in order, amongst them will be found the progression $n+r$, $r$, $-n+r$, &c. therefore, by trial, the divisor $mx^2+nx+r$ will be discovered, if the equation admit of a quadratic divisor whose coefficients are whole numbers.

Let the proposed equation be $3x^4+4x^3+3x^2-2x+2=0$:

| Sup. | Res. | Divisors. | Sq. | Sums and Differences. | Progress. |
|---|---|---|---|---|---|
| $x = 1$ | 10 | 1, 2, 5, 10 | 3 | $-7, -2, 1, 2, 4, 5, 8, 13$ | $-2, 1$ |
| $x = 0$ | 2 | 1, 2 | 0 | $-2, -1, 1, 2$ | $2, 1$ |
| $x = -1$ | 6 | 1, 2, 3, 6 | 3 | $-3, 0, 1, 2, 4, 5, 6, 9$ | $6, 3$ |
| $x = -2$ | 34 | 1, 2, 17, 34 | 12 | $-22, -5, 10, 11, 13, 14, 29, 46$ | $10, 5$ |

From the first progression, $n = -4$, $r = 2$; from the other, $n = 2$, and $r = -1$; therefore, since $m$ may either be positive or negative, the divisors to be tried are

are $\pm 3x^2 - 4x + 2$, and $\pm 3x^2 + 2x - 1$; of which, $- 3x^2$ $+ 2x - 1$, or $3x^2 - 2x + 1$ succeeds; consequently, the roots of the equation $3x^2 - 2x + 1 = 0$, are two roots of the proposed biquadratic.

# THE METHOD OF APPROXIMATION.

(341.) The most useful and general method of discovering the possible roots of numeral equations, is approximation. Find by trial (Art. 318), two numbers, which substituted for the unknown quantity give, one a positive, and the other a negative result; and an odd number of roots lies between these two quantities, that is, one possible root at least, lies between them; then by increasing one of the limits, and diminishing the other, an approximation may be made to the root; substitute this approximate value, increased or diminished by $v$, for the unknown quantity in the equation, neglect all the powers of $v$ above the first, as being small when compared with the other terms, and a simple equation is obtained for determining $v$, nearly; thus a nearer approximation is made to the root, and by repeating the operation, the approximation may be made to any required degree of exactness.

Ex. Let the roots of the equation $y^3 - 3y + 1 = 0$ be required.

When 1 is substituted for $x$ the result is $-1$, and when 2 is substituted, the result is $+3$, therefore one possible root lies between 1 and 2; try 1.5, and the result is $-.125$, or the root lies between 1.5 and 2.

Let

Let $1.5 + v = y$; then,

$$\left.\begin{array}{rl} y^3 = & 3.375 + 6.75\,v + 4.5\,v^2 + v^3 \\ -3y = & -4.5 \quad -3v \\ +\ 1 = & +1 \end{array}\right\} = 0,$$

that is, $-.125 + 3.75\,v + 4.5\,v^2 + v^3 = 0$, and neglecting the two last terms, $-.125 + 3.75\,v = 0$, or $v = \dfrac{.125}{3.75} = .033$ nearly, and $y = 1.5 + v = 1.533$ nearly.

Again, suppose $1.533 + v = y$; by proceeding as before, we find $003686437 + 4.050267\,v = 0$, and $v = \dfrac{-.003686437}{4.050267} = -.0009101$ &c. hence, $y = 1.532089$ nearly. The other roots may be found by the solution of a quadratic (Art. 267).

(342.) *The accuracy of the approximation does not depend upon the ratio which the quantity assumed bears to the root, but upon it's being nearer to one root than to any other.*

Let the roots of the equation $x^n - p\,x^{n-1} + q\,x^{n-2} - $ &c. $= 0$, be $a+m$, $a+n$, $a+r$, &c. of which $a+m$ is the least; assume $a+v=x$, and let $P - Qv + Rv^2 - Sv^3 + $ &c. $=0$ be the transformed equation, whose roots are $m$, $n$, $r$, &c. then $\dfrac{Q}{P} = \dfrac{1}{m} + \dfrac{1}{n} + \dfrac{1}{r} + $ &c. (Art. 273),

and $\dfrac{P}{Q} = \dfrac{1}{\dfrac{1}{m} + \dfrac{1}{n} + \dfrac{1}{r} + \&c.}$ : In the process we assume $P$

$-Qv=0$, or $v = \dfrac{P}{Q} = \dfrac{1}{\dfrac{1}{m} + \dfrac{1}{n} + \dfrac{1}{r} + \&c.}$ ; and on suppo-

sition

sition that $m$ is much less than $n$ or $r$, &c. $\dfrac{1}{m}$ is much

greater than $\dfrac{1}{n} + \dfrac{1}{r} +$ &c. and $\dfrac{P}{Q} = m$ nearly; but if $m$

bear a finite ratio to $n$ or $r$, the approximation will be less accurate, and the less these magnitudes $n$, $r$, &c. are, the greater error is made in supposing $\dfrac{1}{n} + \dfrac{1}{r} +$ &c.

to vanish, when compared with $\dfrac{1}{m}$.

(343.) When $m$ and $n$ are nearly equal to each other, and much less than $r$, $s$, &c. and also both positive or both negative, $\Big\{$ then $\dfrac{P}{Q} = \dfrac{1}{\dfrac{1}{m} + \dfrac{1}{n}} = \dfrac{m}{1 + \dfrac{m}{n}}$, nearly,

which is an approximation to $m$ the less of the two; but if one of these quantities be positive and the other negative, $\dfrac{1}{m} + \dfrac{1}{n}$ may be either positive or negative,

and greater, equal to, or less than $\dfrac{1}{r} + \dfrac{1}{s} +$ &c. and

consequently $\dfrac{P}{Q}$ is not necessarily an approximation to

any of the quantities $m$, $n$, $r$, $s$, &c. $\Big\}$

Let $P - Qv + Rv^2 = 0$; the roots of this equation will be $m$ and $n$, nearly. For if $m$, $n$, $r$, $s$, be the roots of the equation $P - Qv + Rv^2 - Sv^3 +$ &c. $= 0$, $P = mnrs$, $Q = mns + mnr + mrs + nrs$, $R = mn + mr + ms + nr + ns + rs$, and since $m$ and $n$ are small when compared with $r$ and $s$, $Q = mrs + nrs$ nearly, and $R = rs$ nearly; therefore the equation $P - Qv + Rv^2 = 0$ becomes $mnrs - \overline{mrs + nrs} \cdot v + rsv^2 = 0$;

hence,

hence, $v^2 - \overline{m+n}.v + mn = 0$, whose roots are $m$ and $n$. By the solution then of this quadratic, a much nearer approximation is made to the root $a+m$ than by the former method, and at the same time, an approximation is also made to the root $a+n$.

(344.) In the same manner, if $t$ roots be nearly equal, in order to approximate to them, it will be necessary to solve an equation of $t$ dimensions. See Dr. Waring's *Med. Algeb.* p. 186.

(345.) If we have two equations, containing two unknown quantities, we may discover the values of these quantities nearly in the same manner.

## Ex.

Let $\left\{\begin{matrix} x^2 y = 405 \\ xy - y^2 = 20 \end{matrix}\right\}$ To find $x$ and $y$.

Find, by trial, approximate values of $x$ and $y$; such are 20 and 1; and let $x = 20+v$, $y = 1+z$;
then $x^2 y = 400 + 40v + 400z + v^2 + 40vz + v^2z = 405$,
and $xy - y^2 = 19 + v + 18z + vz - z^2 = 20$,
and neglecting those terms in which $z$ or $v$ is of more than one dimension, or in which their product is found, as being small when compared with the rest,

$$400 + 40v + 400z = 405$$
$$19 + v + 18z = 20$$
$$40v + 400z = 5$$
$$v + 10z = .125$$
$$\text{and } v + 18z + 19 = 20$$
$$\text{or } v + 18z = 1$$
$$\text{hence, } 8z = .875$$
$$z = .109375$$

$$v =$$

$$v = .125 - 10z = -.96875$$
therefore $x = 19.03$
and $y = 1.109$.

By making use of the values thus obtained, nearer approximations may be made to $x$ and $y$.

## ON THE REVERSION OF SERIES.

(346.) If two quantities $Ax + Bx^2 + Cx^3 + \&c.$ and $ax + bx^2 + cx^3 + \&c.$ be always equal, the invariable coefficients of the corresponding terms are equal.

For if these equal quantities be divided by $x$, we have $A + Bx + Cx^2 + \&c. = a + bx + cx^2 + \&c.$ or when $x$ vanishes, $A = a$, and $A$ and $a$ are invariable, therefore in all cases, $A = a$; hence also, $Bx + Cx^2 + \&c. = bx + cx^2 + \&c.$ or dividing by $x$, $B + Cx + \&c. = b + cx + \&c.$ and when $x$ vanishes, $B = b$, therefore in all cases, $B = b$. In the same manner, $C = c$, &c.

(347.) Cor. If $A + Bx + Cx^2 + \&c. = 0$ in all cases, then $A = 0$, $B = 0$, $C = 0$, &c.

(348.) Approximation may be made to a root of an equation, by assuming for it a series, involving the powers of that quantity in terms of which it is sought, with indeterminate coefficients; this series being substituted for the unknown quantity in the proposed equation, the coefficients may be found by making each term equal to 0, and thus the series, which expresses the value of the unknown quantity, may be determined.

Ex.

**Ex.** Let $y^3 - 3y + x = 0$; required the value of $y$ in terms of $x$.

Let $y = ax + bx^3 + cx^5 + dx^7 + $ &c. then

$$y^3 = \qquad a^3x^3 + 3\,a^2bx^5 + 3\,a^2cx^7 + \text{&c.}$$
$$+ 3\,ab^2x^7$$
$$-3y = -3\,ax - 3\,bx^3 - \quad 3\,cx^5 - \quad 3\,dx^7 - \text{&c.}$$
$$+ x = + \quad x$$

$\left.\right\} = 0,$

and supposing each term to vanish (Art. 347), $-3\,a$

$+1 = 0$, or $a = \dfrac{1}{3}$; $a^3 - 3\,b = 0$, $b = \dfrac{a^3}{3} = \dfrac{1}{3^4}$; $3\,a^2b - 3\,c$

$= 0$, and $c = a^2b = \dfrac{1}{3^6}$; &c. therefore

$y = \dfrac{x}{3} + \dfrac{x^3}{3^4} + \dfrac{x^5}{3^6} + $ &c. and when $x = 1$, $y = \dfrac{1}{3} + \dfrac{1}{3^4} +$

$\dfrac{1}{3^6} + $ &c. $= .347$ &c. which is one root of the equation

$y^3 - 3y + 1 = 0$.

If for $y$, the series $ax + bx^2 + cx^3 + dx^4 + $ &c. had been assumed, the quantities $b$, $d$, &c. would have been found $= 0$; therefore the even terms are unnecessary.

(349.) Cor. The less $x$ is assumed, the faster will this series converge, and the more accurately will $y$ be obtained.

(350.) This method of approximation is similar to the former, in this respect, that the series will have a slow degree of convergency, unless one value of $y$ be much less than any other. If this be not the case, find $m$, an approximate value of $y$, by trial, and assume $m \pm v = y$; then, when one value of $v$ is much less than any other, it may be found by this reversion,

and

and consequently that value of $y$ which is nearest to $m$, will be known.

In the example Art. 341, $y$ being found nearly 1.5, assume $v + 1.5 = y$, and the equation is transformed into $v^3 + 4.5v^2 + 3.75v - .125 = 0$: call this $v^3 + pv^2 + qv - x = 0$, and take $v = ax + bx^2 + cx^3 + dx^4 + \&c.$ then

$$
\left.
\begin{array}{l}
v^3 = \qquad\qquad a^3x^3 + \qquad 3a^2bx^4 + \&c. \\
+pv^2 = \quad pa^2x^2 + 2pabx^3 + p.\overline{b^2 + 2ac}.x^4 + \&c. \\
+qv = qax + qbx^2 + \quad qcx^3 + \qquad qdx^4 + \&c. \\
-x = -x
\end{array}
\right\} = 0,
$$

hence, $qa - 1 = 0$, and $a = \dfrac{1}{q} = \dfrac{1}{3.75} = .26666$ &c. $qb + pa^2 = 0$, and $b = \dfrac{-pa^2}{q} = -.08533$ &c. therefore $v =$

$.26666 \times .125 - .08533 \times .\overline{125}\,]^2 + \&c. = .0320$ &c. and $y = v + 1.5 = 1.5320$ &c. as before.

(351.) The same method may be used to find $y$ in terms of $x$, when $x$ and $y$, and their powers, are combined in any manner in the equation.

Ex. 1. Let $x = ay + by^2 + cy^3 + \&c.$ required the value of $y$ in terms of $x$.

Assume $y = Ax + Bx^2 + Cx^3 + \&c.$

then 
$$
\left.
\begin{array}{l}
ay = aAx + aBx^2 + \quad aCx^3 + \&c. \\
by^2 = \qquad bA^2x^2 + 2bABx^3 + \&c. \\
cy^3 = \qquad\qquad cA^3x^3 + \&c. \\
\&c. = \qquad\qquad \&c. \\
-x = -x
\end{array}
\right\} = 0,
$$

therefore $aA - 1 = 0$, or $A = \dfrac{1}{a}$; $aB + bA^2 = 0$, or $B =$

$$B = \frac{-bA^2}{a} = \frac{-b}{a^3} \; ; \; aC + 2bAB + cA^3 = 0, \text{ or } C =$$

$$\frac{-2bAB}{a} - \frac{cA^3}{a} = \frac{2b^2 - ac}{a^5}; \; \&c. \text{ hence, } y = \frac{x}{a} - \frac{bx^2}{a^3}$$

$$+ \frac{\overline{2b^2 - ac}.x^3}{a^5} \; \&c.$$

Ex. 2. Let $x = y - ay^3 + by^5 -$ &c. required the value of $y$ in terms of $x$.

Assume $y = Ax + Bx^3 + Cx^5 +$ &c.

$$\left.\begin{array}{l} \text{then } y = Ax + \quad Bx^3 + \quad\quad Cx^5 + \&c. \\ \quad - ay^3 = \quad\quad - aA^3x^3 - 3aA^2Bx^5 - \&c. \\ \quad + by^5 = \quad\quad\quad\quad + \quad bA^5x^5 + \&c. \\ \quad \&c. = \quad\quad\quad\quad\quad \&c. \\ \quad - x = - x \end{array}\right\} = 0,$$

hence, $A - 1 = 0$, or $A = 1$; $B - aA^3 = 0$, or $B = a$; $C - 3aA^2B + bA^5 = 0$, or $C = 3a^2 - b$; &c. therefore $y = x + ax^3 + \overline{3a^2 - b}.x^5 +$ &c.

The method of determining the proper series to be assumed in each case, without previous trial, is given by Maclaurin; *Alg.* Pt. 2. Ch. 10.

# ON THE SUMS OF THE POWERS OF THE ROOTS OF AN EQUATION.

(352.) Let $a$, $b$, $c$, &c. be the roots of the equation $x^n - px^{n-1} + qx^{n-2} \ldots\ldots + wx^{n-m} -$ &c. $= 0$, and $A$, $B$, $C, \ldots\ldots P$, $Q$, $R$, $S$, the sum of the roots, sum of their squares, cubes, $\ldots\ldots m - 3$, $m - 2$, $m - 1$, $m$, powers respectively; then will $A = p$, $B = pA - 2q$, $C = pB - qA + 3r$, &c. and $S = pR - qQ + rP \ldots\ldots$
$$- mw;$$

$- mw$; where $+ w$ is the coefficient of the $\overline{m + 1}^{\text{th}}$ term.

It appears by Art. 308 that

$$\frac{nx^{n-1} - \overline{n-1}.px^{n-2} + \overline{n-2}.qx^{n-3}\ldots + \overline{n-m}.wx^{n-m-1} - \&\text{c.}}{x^n - px^{n-1} + qx^{n-2}\ldots + wx^{n-m} - \&\text{c.}}$$

$$= \frac{1}{x-a} + \frac{1}{x-b} + \frac{1}{x-c}\&\text{c. whatever be the value}$$

of $x$; and by actual division,

$$\frac{1}{x-a} = \frac{1}{x} + \frac{a}{x^2} + \frac{a^2}{x^3}\ldots + \frac{a^m}{x^{m+1}} + \&\text{c.}$$

$$\frac{1}{x-b} = \frac{1}{x} + \frac{b}{x^2} + \frac{b^2}{x^3}\ldots + \frac{b^m}{x^{m+1}} + \&\text{c.}$$

$$\frac{1}{x-c} = \frac{1}{x} + \frac{c}{x^2} + \frac{c^2}{x^3}\ldots + \frac{c^m}{x^{m+1}} + \&\text{c.}$$

&c.

and if $x$ be supposed greater than any of the magnitudes $a$, $b$, $c$, &c. no quantity is lost in the division; therefore by addition,

$$\frac{nx^{n-1} - \overline{n-1}.px^{n-2} + \overline{n-2}.qx^{n-3}\ldots + \overline{n-m}.wx^{n-m-1}\&\text{c.}}{x^n - px^{n-1} + qx^{n-2}\ldots + wx^{n-m} - \&\text{c.}}$$

$$= \frac{n}{x} + \frac{A}{x^2} + \frac{B}{x^3}\ldots + \frac{P}{x^{m-2}} + \frac{Q}{x^{m-1}} + \frac{R}{x^m} + \frac{S}{x^{n+1}} +$$

&c. and multiplying by $x^n - px^{n-1} + qx^{n-2}\ldots + wx^{n-m} - \&\text{c.}$ we have

$$nx^{n-1} - \overline{n-1}.px^{n-2} + \overline{n-2}.qx^{n-3}\ldots + \overline{n-m}.wx^{n-m-1}$$
$$- \&\text{c.} =$$

$$nx^{n-1}$$

$$\left.\begin{array}{l} nx^{n-1} + Ax^{n-2} + Bx^{n-3}\dots + \ Sx^{n-m-1} + \&c. \\ \quad - npx^{n-2} - pAx^{n-3}\dots - pRx^{n-m-1} - \&c. \\ \qquad + nqx^{n-3}\dots + qQx^{n-m-1} + \&c. \\ \qquad\qquad \dots - rPx^{n-m-1} - \&c. \\ \qquad\qquad\qquad \&c. \\ \qquad\qquad\qquad + nwx^{n-m-1} + \&c. \end{array}\right\}$$

and by equating the coefficients, $A - np = -\overline{n-1}.p,$ and $A = p$; $B - pA + nq = \overline{n-2}.q$, or $B = pA - 2q$, &c. $S - pR + qQ - rP\dots + nw = \overline{n-m}.w$, or $S = pR - qQ + rP\dots - mw.$

Ex. Let $x^3 + 5x^2 - 6x - 8 = 0$; then by comparing the terms of this, with the terms of the equation $x^n - px^{n-1} + qx^{n-2} - \&c. = 0$, we have $p = -5$, $q = -6$, $r = 8$.

Hence, the sum of the roots $= p = -5 = A.$
Sum of the squares $= pA - 2q = 25 + 12 = 37 = B.$
Sum of the cubes $= pB - qA + 3r = -185 - 30 + 24 = -191 = C$ &c.

(353.) The proposition also admits of the following proof.

The same notation being retained; let $m$ and $n$ be equal, and since $a$, $b$, $c$, &c. are roots of the equation,

$$a^n - pa^{n-1} + qa^{n-2}\dots + w = 0$$
$$b^n - pb^{n-1} + qb^{n-2}\dots + w = 0$$
$$c^n - pc^{n-1} + qc^{n-2}\dots + w = 0$$
$$\&c.$$

by add. $S - pR + qQ\dots + nw = 0,$
    or $S = pR - qQ\dots - nw.$

If $m$ be greater than $n$; multiply the proposed equation by $x^{m-n}$, then $x^m - px^{m-1} + qx^{m-2}\dots + wx^{m-n} = 0$;

$=0$; which equation has the roots $a$, $b$, $c$, &c. and $m-n$ roots each equal to $o$; therefore the sum of the $m^{th}$ powers of the roots of this equation is equal to the sum of the $m^{th}$ powers of the roots of the former; that is, $S = pR - qQ +$&c. to $n$ terms.

When $m$ is less than $n$: The sum of the $m^{th}$ powers of the roots may be expressed in terms of $p$, $q$, $r$, . . . .$w$, where $w$ is the coefficient of the $m+1^{th}$ term of the equation.    For $p^2$ contains $a^2 + b^2 + c^2 +$ &c. with other combinations of the roots, as $ab$, $ac$, $bc$, &c. which combinations are contained in a multiple of $q$; also, $p^3$ contains $a^3 + b^3 + c^3 +$&c., with other combinations, such as $a^2b$, $a^2c$, $b^2a$, &c. $abc$, $acd$, $bcd$, &c. and these combinations may be made up of $p$, $q$ and $r$; for $p \times q$ contains the quantities $a^2b$, $a^2c$, $b^2a$, &c. and $r$ is the sum of the quantities $abc$, $acd$, $bcd$, &c. In the same manner it appears, that $a^4 + b^4 + c^4 +$ &c. may be found in terms of $p$, $q$, $r$ and $s$; and in general, $a^m + b^m + c^m +$&c. may be expressed in terms of $p$, $q$, $r$, . . . .$w$.    Also, the number of combinations of any particular form, as $a^2b$, cannot be altered by the introduction of the root $c$; consequently the numeral coefficient of the product $pq$, by which the combinations of that form are taken away, is the same, whatever be the number of roots: Hence the expression for $a^m + b^m + c^m$ &c. in the equation $x^n - px^{n-1} + qx^{n-2} . . . . . + wx^{n-m} -$&c. $= 0$, is the same with the expression for the sum of the $m^{th}$ powers of the roots of the equation $x^m - px^{m-1} + qx^{m-2} . . . . + w = 0$; that is, $S = pR - qQ . . . . - mw$.

This

This rule was given by Sir I. Newton for the purpose of approximating to the greatest root of an equation. Suppose the roots all possible, and one greater than the rest, the powers of this root increase in a higher ratio than those of any. other, and the $2m^{th}$ power of this root will approach nearer and nearer to a ratio of equality with the sum of the $2m^{th}$ powers of the roots, as $m$ increases; therefore by extracting the $2m^{th}$ root of this sum, an approximation is made to the greatest positive or least negative root (See Art. 306).

## ON THE IMPOSSIBLE ROOTS OF AN EQUATION.

(354.) It has before been shewn (Art. 311), that there are as many positive roots in an equation as it has changes of signs, and as many negative roots as continuations of the same sign, when the roots are all possible.

But this rule cannot be applied to impossible roots; which appears by the demonstration there given, as well as from the consideration, that an impossible expression cannot be said to be either positive or negative.

If then it appear from the terms of an equation that some roots may, according to the above rule, either be called positive or negative, they must be impossible. Thus, two roots of the equation $x^3 + qx + r = 0$, or $x^3 \pm o + qx + r = 0$, are impossible; because

cause it has two changes of signs or none, according as the second term is supposed to be $-o$, or $+o$. In the same manner, if any term of an equation be wanting, and the signs of the adjacent terms be both positive or both negative, the equation has, at least, two impossible roots: and if two succeeding terms be wanting, it must always have, at least, two impossible roots.

(355.) Impossible roots enter equations by pairs (Art. 277); they also lie under the form of two positive or two negative roots.

Let $\pm a + \sqrt{-b^2}$ and $\pm a - \sqrt{-b^2}$ be the roots; then $\overline{x \mp a - \sqrt{-b^2}} \times \overline{x \mp a + \sqrt{-b^2}} = x^2 \mp 2ax + a^2 + b^2 = 0$, the *signs* of which equation shew that it has either two positive, or two negative roots.

(356.) Cor. Hence, if the last term of an equation of an even number of dimensions be negative, it will have at least two possible roots, one positive and the other negative (Art. 271).

(357.) Let an equation be transformed into one whose roots are the squares of the roots of the former, (Art. 296), then as many negative roots as the transformed equation contains, so many impossible roots, at least, are in the original equation, because the square of a possible quantity is always positive.

(358.) *If any series of magnitudes be substituted in order, for the unknown quantity in an equation, there can only be as many changes of signs in the results, as the equation contains possible roots.*

Let

Let $\overline{x^2 - 2ax + a^2 + b^2} \times \overline{x - c}.\overline{x - d}$. &c. $= 0$, be an equation whose roots are $a + \sqrt{-b^2}$, $a - \sqrt{-b^2}$, $c$, $d$, &c. whatever magnitude is substituted for $x$, the quantity $x^2 - 2ax + a^2 + b^2$, which is the sum of two squares, $\overline{x - a}\rvert^2 + b^2$, is positive; therefore the changes of signs can only arise from the substitution of quantities for $x$ in the product $\overline{x - c}.\overline{x - d}$. &c. which changes, when the quantities are taken in order, are as many as there are possible roots $c$, $d$, &c. (See Art. 298.)

(359.) *The limiting equation\* has, at least, as many possible roots as the original equation, wanting one.*

Let $\overline{x^2 - 2ax + a^2 + b^2} \times \overline{x - c}.\overline{x - d}$. &c. $= 0$ be the proposed equation, the limiting equation is

$$\left. \begin{array}{l} \overline{x^2 - 2ax + a^2 + b^2} \times \overline{x - c}. \text{ &c.} \\ + \overline{x^2 - 2ax + a^2 + b^2} \times \overline{x - d}. \text{ &c.} \\ + \overline{x - a + \sqrt{-b^2}} \times \overline{x - c}.\overline{x - d}. \text{ &c.} \\ + \overline{x - a - \sqrt{-b^2}} \times \overline{x - c}.\overline{x - d}. \text{ &c.} \end{array} \right\} = 0,$$

or by adding the two last terms together, it is

$$\left. \begin{array}{l} \overline{x^2 - 2ax + a^2 + b^2} \times \overline{x - c}. \text{ &c.} \\ + \overline{x^2 - 2ax + a^2 + b^2} \times \overline{x - d}. \text{ &c.} \\ + \quad 2.\overline{x - a}.\overline{x - c}. \quad \overline{x - d}. \text{ &c.} \end{array} \right\} = 0,$$

(See Art. 307), in which, if $c$, $d$, &c. be successively substituted for $x$, the results are $+$, $-$, &c. therefore there are possible roots in this latter equation which lie between $c$, $d$, &c. or this equation contains at least, as many possible roots, wanting one, as the original equation. It may contain more.

(360.) Cor.

---

\* This is the limiting equation mentioned Art. 307; the proposition is not necessarily true of the other limiting equations, (Art. 314.)

(360.) Cor. 1. Hence it follows, that there are, at least, as many impossible roots in the original equation as in the equation of limits. There may be more; therefore from the number of impossible roots in the limiting equation, we cannot determine, exactly, the number in the original equation.

(361.) Cor. 2. Hence also it appears, that if the possible roots of the limiting equation be substituted successively in the original equation, we know from the signs of the results, what possible roots the latter contains. For, roots of the limiting equation lie between the possible roots of the proposed equation (Art. 359.)

Ex. Let $x^{n+3} - a^{n+1}x^2 + p^2 d^{n+1} = 0$.

Its limiting equation is $\overline{n+3}.x^{n+2} - 2a^{n+1}x = 0$; whose possible roots, when $n$ is an odd number, are $-\overline{\dfrac{2}{n+3}}\Big|^{\frac{1}{n+1}} \times a$;

$o$, and $\overline{\dfrac{2}{n+3}}\Big|^{\frac{1}{n+1}} \times a$; which substituted in the original equation give the results either $-$, $+$, $-$, or $+$, $+$, $+$; therefore it has either four possible roots or none. When $n$ is even, the possible roots of the limiting equation are $o$ and $\overline{\dfrac{2}{n+3}}\Big|^{\frac{1}{n+1}} \times a$; therefore the equation itself will have one possible root or three, according as $\overline{\dfrac{2}{n+3}}\Big|^{\frac{1}{n+1}} \times a$, when substituted for $x$, gives a positive or a negative result.

(362.) The roots of a quadratic equation are impossible, if the square of the middle term be less than four times the product of the extremes.

Let $ax^2 + bx + c = 0$; then $x = \dfrac{-b \pm \sqrt{b - 4ac}}{2a}$,

which

which expression becomes impossible when $b^2$ is less than $4ac$.

(363.) It appears from Art. 360, that there are impossible roots in an equation, whenever there are impossible roots in it's limiting equation. In the same manner, if the next limit be taken, there are impossible roots in the original equation, whenever there are impossible roots in this limit; and if the limit be thus brought down to a quadratic, when the roots of the quadratic are impossible, there are impossible roots in the original equation corresponding to them. On this principle is founded Sir I. Newton's rule for discovering impossible roots in any equation.

Let the proposed equation be $x^n - p x^{n-1} \ldots + D x^{n-r+1} - E x^{n-r} + F x^{n-r-1} - \&c. = 0$. To obtain a limiting equation, which shall be a quadratic, corresponding to the terms $D x^{n-r+1} - E x^{n-r} + F x^{n-r-1}$, let the succeeding terms be taken away, by multiplying by the terms of the arithmetical progressions $n, n-1, n-2, \ldots 2, 1, 0; \ n-1, n-2, \ldots 2, 1, 0; \ n-2, n-3, \ldots 2, 1, 0; \ \&c.$ and let the preceding terms be taken away, by multiplying by the terms of the progressions, $0, 1, 2, \ldots r+1; \ 0, 1, 2, \ldots r; \ \&c.$ as follows;

$$-px^{n-1} + qx^{n-2}\ldots\ldots + Dx^{n-r+1} - Ex^{n-r} + Fx^{n-r-1}\ldots\ldots = 0$$

| $x^n$ | | | | | | | | | |
|---|---|---|---|---|---|---|---|---|---|
| $n,$ | $n-1,$ | $n-2,$ | $n-r+1,$ | $n-r,$ | $n-r-1,$ | | | | |
| $n-1,$ | $n-2,$ | $n-3,$ | $n-r,$ | $n-r-1,$ | $n-r-2,$ | | | | |
| $n-2,$ | $n-3,$ | $n-4,$ | $n-r-1,$ | $n-r-2,$ | $n-r-2,n-r-3,$ | | | | |
| &c. | | | &c. | | | | | | |
| | | | $4,$ | $3,$ | $2,$ | | | | |
| | | | $3,$ | $2,$ | $1,$ | | | | |
| | $2,$ | | $r-1,$ | $r,$ | $r+1,$ | | | | |
| | $1,$ | | $r-2,$ | $r-1,$ | $r,$ | | | | |
| $0,$ | $0,$ | | $r-3,$ | $r-2,$ | $r-1,$ | | | | |
| | | | &c. | | | | | | |
| | | | $2,$ | $3,$ | $4,$ | | | | |
| | | | $1,$ | $2,$ | $3,$ | | | | |
| $*$ | $*$ | | $+\overline{n-r+1}.\overline{n-r}.Dx^2 - 2.\overline{n-r}.rEx + \overline{r+1}.rF$ | | $* = 0$ | | | | |

The respective products being taken, and those quantities left out which are found in every product, we obtain a limiting quadratic $\overline{n-r+1}.\overline{n-r}.Dx^2 - 2.\overline{n-r}.rEx + \overline{r+1}.rF = 0$, corresponding to the three terms $Dx^{n-r+1} - Ex^{n-r} + Fx^{n-r-1}$; and if two roots of this quadratic be impossible, that is, if $\overline{n-r}|^2 \times r^2E^2$ be less

less than $\overline{n-r+1}\,.\,\overline{n-r}\,.\,\overline{r+1}\,.\,r\,DF$ (Art. 362), or

$\dfrac{\overline{n-r}.r}{n-r+1.r+1}\,E^2$ less than $DF$, there are impossible

roots in the proposed equation, corresponding to them*.

Write down therefore a series of fractions $\dfrac{n}{1}$, $\dfrac{n-1}{2}$,

$\dfrac{n-2}{3}$, .... $\dfrac{n-r+1}{r}$, $\dfrac{n-r}{r+1}$; divide each fraction by

that which precedes it, and place the results, $\dfrac{n-1}{2n}$,

$\dfrac{\overline{n-2}.2}{n-1.3}$, .... $\dfrac{\overline{n-r}.r}{n-r+1.r+1}$, over the succeeding terms

of the equation, beginning with the second; also, place the sign $+$ under the first and last terms, and under every other term $+$ or $-$, according as the square of that term multiplied by the fraction which stands over it, is greater or less than the product of the adjacent terms; then there will be as many impossible roots in the equation as there are changes of these signs from $+$ to $-$ and from $-$ to $+$.

Ex. Let the proposed equation be $x^3 - 4x^2 + 4x - 6 = 0$.

In the series of fractions $\dfrac{3}{1}$, $\dfrac{2}{2}$, $\dfrac{1}{3}$, if each term be

divided by that which precedes it, we obtain $\dfrac{1}{3}$, $\dfrac{1}{3}$, to

be placed over the terms of the equation;

$$\overset{\tfrac{1}{3}\qquad\ \tfrac{1}{3}}{x^3 - 4x^2 + 4x - 6 = 0}$$
$$+\quad +\quad -\quad +$$

and

---

and since $\dfrac{\overline{-4}\,|^2}{3}$, or $\dfrac{16}{3}$, is greater than $4 \times 1$, the sign $+$ must be placed under the second term; but $\dfrac{4^2}{3}$, or $\dfrac{16}{3}$, is less than $-4 \times -6$, or $24$, therefore the sign $-$ must be placed under the third term; and $+$ being placed under the first and last terms, there are two changes of signs; therefore the equation contains two impossible roots.

## SCHOLIUM.

(364.) The discovery of the number of impossible roots in an equation has given great trouble to Algebraists, and their researches, hitherto, have not been attended with any great success. In a cubic equation $x^3 - qx + r = 0$ two roots are impossible or not, according as $\dfrac{r^2}{4} - \dfrac{q^3}{27}$ is positive or negative (Art. 331). A biquadratic, $x^4 + qx^2 + rx + s = 0$, has two impossible roots, when two roots of the equation $y^3 + 2qy^2 + \overline{q^2 - 4s}.y - r^2 = 0$, are impossible; and all it's roots are impossible, when the roots of this cubic are all possible and two of them negative (Art. 333).

(365.) Dr. Waring has given a rule for determining the number of impossible roots in an equation of five dimensions, but the investigation cannot properly be introduced into an elementary treatise. See *Med. Algebraicæ*, p. 82.

(366.) Sir

(366.) Sir I. Newton's rule (Art. 363) is general and easily applied, but as it is deduced from the nature of the inferior limits, it will not always detect impossible roots (Art. 360). The proof also is defective, as it does not extend to that part of the rule which respects the *number* of impossible roots. Thus far however it may be depended upon, that it never shews impossible roots, but when there are some such in the proposed equation.

Many other rules, which will frequently discover the impossible roots in any equation, may be seen in the *Med. Alg.* C. 2.

### THE END OF PART II.

THE

THE

# ELEMENTS OF ALGEBRA.

## PART III.

### ON UNLIMITED PROBLEMS.

(367.) WHEN there are more unknown quantities than independent equations, the number of corresponding values which those quantities admit, is indefinite (Art. 145). This number may be lessened, by rejecting all the values which are not integers; it may be farther lessened, by rejecting all the negative values; and still farther, by rejecting all values which are not square, or cube numbers; &c. By restrictions of this kind, the number of answers may be confined within definite limits.

(368.) *If a simple equation express the relation of two unknown quantities, and their corresponding integral values be required; divide the whole equation by the coefficient which is the less of the two, and suppose that part of the quotient, which is in a fractional form, equal to some whole number; thus a new*

simple

*simple equation is obtained, with which we may pro-*
*ceed as before; let the operation be repeated, till the*
*coefficient of one of the unknown quantities is unity,*
*and the coefficient of the other a whole number ; then*
*an integral value of the former may be obtained by*
*substituting* 0, *or any whole number for the other ;*
*and from the preceding equations, integral values of*
*the quantities proposed may be found.*

### Ex. 1.

Let $5x + 7y = 29$; to find the corresponding integral values of $x$ and $y$.

Dividing the whole equation by 5, the less coefficient,

$$x + y + \frac{2y}{5} = 5 + \frac{4}{5}$$

$$\text{or } x = 5 - y + \frac{4 - 2y}{5}.$$

Assume $\dfrac{4 - 2y}{5} = p$, or $4 - 2y = 5p$

then $2 - y = 2p + \dfrac{p}{2}$

$$y = 2 - 2p - \frac{p}{2}$$

let $p = 2s$, then $y = 2 - 5s$, and $x = 5 - y + p$
$= 3 + 5s + 2s = 3 + 7s$.

If $s = 0$, then $x = 3$ and $y = 2$, the only positive whole
numbers which answer the condition of the equation ;
for if $s = 1$, then $x = 10$, and $y = -3$ ; and if $s = -1$,
then $x = -4$, and $y = 7$.

Ex. 2.

## Ex. 2.

To find a number which being divided by 3, 4, 5, the remainders are 2, 3, 4, respectively.

Let $x$ be the number,

then $\dfrac{x-2}{3} = p$, a whole number,

or $x = 3p + 2$ ; also, from the second condition,

$\dfrac{x-3}{4}$, or $\dfrac{3p-1}{4} = q$, a whole number,

that is, $3p - 1 = 4q$

or $p = q + \dfrac{q+1}{3}$ ;

let $\dfrac{q+1}{3} = r$, or $q = 3r - 1$,

then $p = 4r - 1$, and $x = 3p + 2 = 12r - 1$ ;

again, from the third condition, $\dfrac{x-4}{5}$ or $\dfrac{12r-5}{5}$ is

a whole number, that is, $2r + \dfrac{2r}{5} - 1$ is a whole num-

ber, therefore $\dfrac{2r}{5}$ is a whole number; let $\dfrac{2r}{5} = 2m$ ;

then $r = 5m$, and $x = 12r - 1 = 60m - 1$ ; if $m = 1$, $x = 59$; if $m = 2$, $x = 119$; &c.

(369.) If the simple equation contain more unknown quantities, their corresponding integral values may be found in the same manner.

## Ex. 3.

Let $4x + 3y + 10 = 5v$ : to find corresponding integral values of $x$, $y$ and $v$.

Dividing

Dividing the whole equation by 3, the least coefficient,

$$x + y + 3 + \frac{x+1}{3} = v + \frac{2v}{3}$$

$$y = v - x - 3 + \frac{2v - x - 1}{3}.$$

Assume $\dfrac{2v - x - 1}{3} = p$, or $2v - x - 1 = 3p$,

then $x = 2v - 3p - 1$

and $y = v - 2v + 3p + 1 - 3 + p = 4p - v - 2$,

and substituting for $p$ and $v$, nothing, or any whole numbers, integral values of $x$ and $y$ are obtained : If $v = 3$, and $p = 1$, then $x = 2$ and $y = -1$; if $v = 4$, and $p = 0$, then $x = 7$ and $y = -6$; &c.

(370.) In the solution of different kinds of unlimited problems, different expedients must be made use of, which expedients, and their application, are chiefly to be learned by practice.

### Ex. 1.

To find what numbers are divisible by 3 without remainders.

Let $a$, $b$, $c$, $d$, &c. be the digits, or figures in the unit's, ten's, hundred's, thousand's, &c. place of any number, then the number is $a + 10b + 100c +$ $1000d +$ &c. this divided by 3 is $\dfrac{a}{3} + 3b + \dfrac{b}{3} + 33c +$ $\dfrac{c}{3} + 333d + \dfrac{d}{3} +$ &c. or $\dfrac{a + b + c + d + \text{&c.}}{3} + 3b + 33c$ $+ 333d +$ &c. which is a whole number when $\dfrac{a + b + c + d + \text{&c.}}{3}$ is a whole number; that is, any

number

number is a multiple of 3 if the sum of it's digits be a multiple of 3. Thus 111, 252, 7851, &c. are multiples of 3.

In the same manner, any number is a multiple of 9 if the sum of it's digits be a multiple of 9.

For $\dfrac{a+10\,b+100\,c+1000\,d+\&c.}{9} = \dfrac{a}{9}+b+\dfrac{b}{9}+11\,c$

$+\dfrac{c}{9}+111\,d+\dfrac{d}{9}+\&c. = \dfrac{a+b+c+d+\&c.}{9}+b+11\,c$

$+\,111\,d+\&c.$ which is a whole number when $\dfrac{a+b+c+d+\&c.}{9}$ is a whole number. Thus 684,6588, &c. are multiples of 9.

Cor. 1. Hence, if any number, and the sum of it's digits, be respectively divided by 9, the remainders are equal.

Cor. 2. From this property of 9 may be deduced a rule which will sometimes detect an error in the multiplication of two numbers.

Let $9\,a+x$ be the multiplicand,

$9\,b+y$ the multiplier,

then $\overline{81\,ab+9\,bx+9\,ay+xy}$ is the product;

and if the sum of the digits in the multiplicand be divided by 9, the remainder is $x$; if the sum of the digits in the multiplier be divided by 9, the remainder is $y$; and if the sum of the digits in the product be divided by 9, the remainder is the same as when the sum of the digits in $xy$ is divided by 9, if there be no mistake in the operation.

Ex. 2.

## Ex. 2.

To find a perfect number, that is, one which is equal to the sum of all the numbers which divide it without remainder.

Suppose $y^n x$ to be a perfect number; it's divisors are $1, y, y^2 \ldots y^n, x, xy, xy^2 \ldots xy^{n-1}$;

therefore $y^n x = 1 + y + y^2 \ldots + y^n + x + xy + xy^2 \ldots + xy^{n-1}$.

Now $1 + y + y^2 \ldots \ldots + y^n = \dfrac{y^{n+1} - 1}{y - 1}$

and $x + xy + xy^2 \ldots \ldots + xy^{n-1} = \dfrac{y^n - 1}{y - 1} \times x$ (Art. 222);

therefore $y^n x = \dfrac{y^{n+1} - 1 + \overline{y^n - 1} \times x}{y - 1}$

or $y^{n+1} x - y^n x = y^{n+1} - 1 + y^n x - x$;

hence, $x = \dfrac{y^{n+1} - 1}{y^{n+1} - 2y^n + 1}$; and that $x$ may be a whole number, let $y^{n+1} - 2y^n = 0$, or $y - 2 = 0$, that is, $y = 2$; then $x = 2^{n+1} - 1$. Also, let $n$ be so assumed that $2^{n+1} - 1$ may have no divisor but unity, which was supposed in taking the divisors of $y^n x$; then $y^n x$, or $2^n \times \overline{2^{n+1} - 1}$ is a perfect number. Thus, if $n = 1$, the number is $2 \times 3$ or $6$, which is equal to $1 + 2 + 3$ the sum of it's divisors: If $n = 2$, the number is $2^2 \times \overline{2^3 - 1} = 4 \times 7 = 28$.

## Ex. 3.

To find two square numbers, whose sum is a square.
Let $x^2$ and $y^2$ be the two squares;

Assume $x^2 + y^2 = \overline{nx - y}\,|^2 = n^2 x^2 - 2 nxy + y^2$,

then

then $x^2 = n^2x^2 - 2nxy$

$x = n^2x - 2ny$

hence, $\overline{n^2 - 1}.x = 2ny$

or $x = \dfrac{2ny}{n^2 - 1}$ .

And if $n$ and $y$ be assumed at pleasure, such a value of $x$ is obtained, that $x^2 + y^2$ is a square number.

But if it be required to find integers of this description, let $y = n^2 - 1$, then $x = 2n$, and $n$ being taken at pleasure, integral values of $x$ and $y$, and consequently of $x^2$ and $y^2$, will be found. Thus, if $n = 2$, then $y = 3$ and $x = 4$, and the two squares are 9 and 16, whose sum is 25.

### Ex. 4.

To find two square numbers, whose difference is a square.

Let $x^2$ and $y^2$ be the two squares;

Assume $x^2 - y^2 = \overline{x - ny}|^2 = x^2 - 2nxy + n^2y^2$

then $-y^2 = -2nxy + n^2y^2$

or $2nx = \overline{n^2 + 1}\ y$

hence, $x = \dfrac{\overline{n^2 + 1}.y}{2n}$

And if $y = 2n$, then $x = n^2 + 1$. Thus, if $n = 2$, then $y = 4$, and $x = 5$; hence, $x^2 - y^2 = 25 - 16 = 9$*.

## ON CONTINUED FRACTIONS.

(371.) To represent $\dfrac{a}{b}$ in a continued fraction.

Let

---

On this subject, see the Edinburgh Transactions, Vol. II. p. 193.

$$b)a(p$$
$$c)b(q$$
$$\overline{\phantom{xxx}}$$
$$d)c(r$$
$$\overline{\phantom{xxx}}$$
$$e \ \&c.$$

Let $b$ be contained in $a$, $p$ times, with a remainder $c$; again, let $c$ be contained in $b$, $q$ times, with a remainder $d$, and so on; then we have

$$a = pb + c$$
$$b = qc + d$$
$$c = rd + e$$
$$\&c.$$

or $\dfrac{a}{b} = p + \dfrac{c}{b} = p + \dfrac{c}{qc + d}$

$$= p + \cfrac{1}{q + \cfrac{d}{c}}$$

$$= p + \cfrac{1}{q + \cfrac{d}{rd + e}}$$

$$= p + \cfrac{1}{q + \cfrac{1}{r + \cfrac{e}{d}}}$$

$$\&c.$$

that is, $\dfrac{a}{b} = p + \cfrac{1}{q + \cfrac{1}{r + \cfrac{1}{s + \&c.}}}$

(372.) Cor. 1. An approximation may thus be made to the value of a fraction whose numerator and denominator are in too high terms, and the farther the division is continued, the nearer will the approximation be to the true value.

(373.) Cor.

(373.) Cor. 2. This approximation is alternately less and greater than the true value. Thus $p$ is less than $\frac{a}{b}$; and $p + \frac{1}{q}$ is greater, because a part of the denominator of the fraction is omitted: $q + \frac{1}{r}$ is too great for the denominator, therefore $p + \dfrac{1}{q + \frac{1}{r}}$ is less than $\frac{a}{b}$; &c.

Ex. To find a fraction which shall be nearly equal to $\frac{314159}{100000}$, and in lower terms.

$$100000)314159(3$$
$$300000$$
$$14159)100000(7$$
$$99113$$
$$887)14159(15$$
$$887$$
$$5289$$
$$4435$$
$$854)887(1$$
$$854$$
$$33\cdot \&c.$$

Here, $p=3$, $q=7$, $r=15$, $s=1$, &c. therefore $\dfrac{314159}{100000}$

$$= 3 + \cfrac{1}{7 + \cfrac{1}{15 + \&c.}}$$

The

The first approximation is 3, which is too little;

the next is $3 + \dfrac{1}{7} = \dfrac{22}{7}$, too great; the next is $3 + \dfrac{1}{7 + \dfrac{1}{15}}$

$= 3 + \dfrac{15}{106} = \dfrac{333}{106}$, too little; and so on. This fraction expresses, nearly, the circumference of a circle whose diameter is unity; therefore the circumference is greater than 3 diameters, less than $\dfrac{22}{7}$ diameters, and greater than $\dfrac{333}{106}$, &c.

(374.) To find the value of a continued fraction, when the denominators $q$, $r$, $s$, &c. recur in any certain order.

Ex. 1. Let $\dfrac{1}{q + \dfrac{1}{r + \dfrac{1}{q + \dfrac{1}{r + \&c.}}}} = x$; then $\dfrac{1}{q + \dfrac{1}{r + x}}$ in inf.

$= x$, or $\dfrac{r + x}{qr + qx + 1} = x$; hence $r + x = qrx + qx^2 + x$,

and $x^2 + rx - \dfrac{r}{q} = 0$, from the solution of which quadratic the value of $x$ may be obtained.

Ex. 2. Let $x = \sqrt{a + \dfrac{b}{\sqrt{a + \dfrac{b}{\sqrt{a + \&c.}}}}}$; by squar-

ing

ing both sides, $x^2 = a + \dfrac{b}{\sqrt{a + \dfrac{b}{\sqrt{a + \&c.}}}} = a + \dfrac{b}{x}$ ; and

$x^3 - ax - b = 0$ ; whence the value of $x$ may be found.

(375.) In the same manner may the values of other quantities, which run on in infinitum, be found, if the factors recur.

Ex. 1. To find the value of $\sqrt{a\sqrt{a\sqrt{a\sqrt{a\,\&c.}}}}$ in infinitum.

Let $\sqrt{a\sqrt{a\sqrt{a\,\&c.}}} = x$ ; by squaring both sides ; $a\sqrt{a\sqrt{a\,\&c.}} = x^2$, that is, $ax = x^2$, or $x = a$.

Ex. 2. Required the value of

$\sqrt{a + \sqrt{b + \sqrt{a + \sqrt{b + \&c.}}}}$ in infinitum.

Let $\sqrt{a + \sqrt{b + \sqrt{a + \sqrt{b + \&c.}}}} = x$ ; by squaring both sides, $a + \sqrt{b + \sqrt{a + \sqrt{b + \&c.}}} = x^2$, and

$\sqrt{b + \sqrt{a + \sqrt{b + \&c.}}} = x^2 - a$ ; therefore

$b + \sqrt{a + \sqrt{b + \&c.}} = x^4 - 2ax^2 + a^2$, or $b + x = x^4 - 2ax^2 + a^2$ ; hence, $x^4 - 2ax^2 - x + a^2 - b = 0$ ; from which equation $x$ may be found.

(376.) *To find the value of a fraction when the numerator and denominator are evanescent.*

Since the value of a fraction depends, not upon the absolute, but the relative magnitude of the numerator and denominator, if in their evanescent state they have a finite ratio, the value of the fraction will be finite. To determine this value, substitute for the

the variable quantity, it's magnitude, when the numerator and denominator vanish, increased by another variable quantity ; then suppose this latter to decrease without limit, and the value of the proposed fraction will be known.

Ex. 1. Required the value of $\dfrac{x^2 - a^2}{x - a}$, when $x = a$.

Let $x = a + v$, and the fraction becomes
$$\frac{a^2 + 2av + v^2 - a^2}{a + v - a} = \frac{2av + v^2}{v} = 2a + v,\ \text{and when}\ v = 0,$$
or $x = a$, the value is $2a$.

Ex. 2. Required the value of $\dfrac{1 - \overline{n + 1} \cdot x^n + n x^{n+1}}{\overline{1 - x}\big|^2}$

when $x = 1$.

Let $x = 1 + v$, and the fraction becomes
$$\frac{1 - \overline{n+1} \cdot \overline{1 + v}\big|^n + n \cdot \overline{1 + v}\big|^{n+1}}{v^2} =$$

$$\frac{1 - \overline{n+1} \times 1 + nv + n \cdot \dfrac{n-1}{2} v^2 + n \cdot \dfrac{n-1}{2} \cdot \dfrac{n-2}{3} v^3 + \&\text{c.}}{v^2} +$$

$$\frac{n \times 1 + \overline{n+1} \cdot v + \overline{n+1} \cdot \dfrac{n}{2} v^2 + \overline{n+1} \cdot \dfrac{n}{2} \cdot \dfrac{n-1}{3} v^3 + \&\text{c.}}{v^2} =$$

$$n \cdot \frac{n+1}{2} + \frac{\overline{n+1} \cdot n \cdot \overline{n-1}}{3} v + \&\text{c. and when}\ v = 0,\ \text{or}$$

$x = 1$, the fraction becomes $n \cdot \dfrac{n+1}{2}$.

(377.) *To find the least common multiple of two quantities; or the least quantity which is divisible by each of them without remainder.*

The

The product of the two quantities divided by their greatest common measure, is their least common multiple.

Let $a$ and $b$ be the two quantities, $x$ their greatest common measure, $m$ their least common multiple, and let $m$ contain $a$, $p$ times, and $b$, $q$ times; that is, let $m = pa = qb$; then $\frac{a}{b} = \frac{q}{p}$; and since $m$ is the least possible, $p$ and $q$ are the least possible; therefore $\frac{q}{p}$ is the fraction $\frac{a}{b}$ in it's lowest terms, and consequently $q = \frac{a}{x}$; hence, $m = qb = \frac{ab}{x}$.

Ex. What is the least common multiple of 18 and 12?

Their greatest common measure is 6; therefore their least common multiple is $\frac{12 \times 18}{6} = 36$.

(378.) Every other common multiple of $a$ and $b$ is a multiple of $m$.

Let $n$ be any other common multiple of the two quantities; and, if possible, let $m$ be contained in $n$, $r$ times, with a remainder $s$, which is less than $m$; then $n - rm = s$; and since $a$ and $b$ measure $n$ and $rm$, they measure $n - rm$, or $s$ (Art. 91); that is, they have a common multiple less than $m$, which is contrary to the supposition.

(379.) *To find the least common multiple of three quantities* a, b *and* c, *take* m *the least common multiple*
*tiple*

*tiple of* a *and* b, *and* n *the least common multiple of* m *and* c; *then* n *is the least common multiple sought.*

For every common multiple of $a$ and $b$ is a multiple of $m$ (Art. 378); therefore every common multiple of $a$, $b$, and $c$ is a multiple of $m$ and $c$; also, every multiple of $m$ and $c$ is a multiple of $a$, $b$ and $c$; consequently the least common multiple of $m$ and $c$ is the least common multiple of $a$, $b$ and $c$.

(380.) Three quantities are said to be in *harmonical proportion,* when the first is to the third, as the difference of the first and second is to the difference of the second and third.

Any magnitudes $A$, $B$, $C$, $D$, $E$, &c. are said to be in *harmonical progression,* if $A : C :: A-B : B-C$; $B : D :: B-C : C-D$; $C : E :: C-D : D-E$; &c.

(381.) *The reciprocals of quantities in harmonical progression are in arithmetical progression.*

Let $A$, $B$, $C$, &c. be in harmonical progression; then $A : C :: A-B : B-C$; therefore $AB - AC = AC - BC$, and dividing both sides by $ABC$,

$$\frac{1}{C} - \frac{1}{B} = \frac{1}{B} - \frac{1}{A}.$$

Again, $B : D :: B-C : C-D$; therefore $BC - BD = DB - DC$, and dividing by $BCD$, $\frac{1}{D} - \frac{1}{C} = \frac{1}{C} - \frac{1}{B}$;

and $\frac{1}{C} - \frac{1}{B}$ has been proved equal to $\frac{1}{B} - \frac{1}{A}$; therefore

the quantities $\frac{1}{A}$, $\frac{1}{B}$, $\frac{1}{C}$, $\frac{1}{D}$, &c. have a common difference, that is, they are in arithmetical progression.

Required

(382.) Required the cube root of the binomial $a + \sqrt{-b^2}$.

Assume $\sqrt[3]{a + \sqrt{-b^2}} = x + \sqrt{-y^2}$;

then $\sqrt[3]{a - \sqrt{-b^2}} = x - \sqrt{-y^2}$ (Art. 257);

by mult. $\sqrt[3]{a^2 + b^2} = x^2 + y^2 = m$, by substitution; therefore $y^2 = m - x^2$; also from the first equation, $\overline{a + \sqrt{-b^2}} = \overline{x + \sqrt{-y^2}}\Big|^3 = x^3 + 3x^2\sqrt{-y^2} - 3xy^2 - y^2\sqrt{-y^2}$; therefore by Art. 253, $x^3 - 3xy^2 = a$; or substituting for $y^2$ it's value $m - x^2$, $x^3 - 3mx + 3x^3 = a$, that is, $4x^3 - 3mx - a = 0$, a cubic equation, whose roots, which are all possible, may be found by approximation, or by a method which will be given in a following part of the work (Art. 515); hence, $y$, and consequently $x + \sqrt{-y^2}$, the root required, may be determined.

In the same manner it appears, that the $c^{\text{th}}$ root may be extracted by the solution of an equation of $c$ dimensions.

# ON LOGARITHMS.

( (383.) If there be a series of magnitudes $a^0$, $a^1$, $a^2$, $a^3$, $\ldots a^x$; $a^{-1}$, $a^{-2}$, $a^{-3}$, $\ldots a^{-y}$, the indices, $0, 1, 2, 3$, $\ldots x$; $-1, -2, -3, \ldots -y$, are called the measures of the ratios of those magnitudes to 1, or the *logarithms* of the magnitudes, for the reason assigned Art. 165.) Thus, $x$, the logarithm of any number $c$, is such a quantity, that $a^x = c$.

Here $a$ may be assumed at pleasure; and for every different value so assumed, a different system of logarithms

rithms will be formed.   In the common tabular logarithms, $a$ is 10, and consequently 0, 1, 2, 3,. . . . $x$,
are the logarithms of 1, 10, 100, 1000,. . . . . $\overline{10}|^{x}$.

(384.)   Cor. 1.   Since the tabular logarithm of 10
is 1, the logarithm of a number between 1 and 10 is
less than 1 ; and in the same manner, the logarithm
of a number between 10 and 100, is between 1 and 2 ;
of a number between 100 and 1000, is between 2
and 3 ; &c.

These logarithms are also real quantities, to which
approximation, sufficiently accurate for all practical
purposes, may be made.

Thus, if $x$ be the logarithm of 5, then $\overline{10}|^{x} = 5$ ;

let $\dfrac{2}{3}$ be substituted for $x$, and $10^{\frac{2}{3}}$ is found to be less

than 5, therefore $\dfrac{2}{3}$ is less than the logarithm of 5 ;

but $10^{\frac{3}{4}}$ is greater than 5, or $\dfrac{3}{4}$ is greater than the

logarithm of 5 ; thus it appears that there is a value

of $x$ between $\dfrac{2}{3}$ and $\dfrac{3}{4}$, such that $\overline{10}|^{x} = 5$ ;  the value

set down in the tables is .69897, and $\overline{10}|^{.69897} = 5$,
nearly.

(385.)   Cor. 2.   If quantities be in geometrical progression, $a^{x}, a^{2x}, a^{3x}$, &c. their logarithms, $x, 2x, 3x$, &c.
are in arithmetical progression.

The method of finding the logarithms of the natural
numbers, or forming a table, is explained in the
Doctrine of Fluxions.

(386.)   *The sum of the logarithms of two numbers
is the logarithm of their product ; and the difference
of the logarithms is the logarithm of their quotient.*

Let

Let $x = $ log. of $c$, and $y = $ log. of $d$; then $a^x = c$ and $a^y = d$; hence, $a^{x+y} = dc$, and $a^{x-y} = \dfrac{c}{d}$; or $x + y$ is the log. of $dc$, and $x - y$ the log. of $\dfrac{c}{d}$.

Ex. 1. Log. of $3 \times 7 = $ log. of $3 + $ log. of $7$.

Ex. 2. Log. of $pqr = $ log. of $pq + $ log. of $r = $ log. of $p + $ log. of $q + $ log. of $r$.

Ex. 3. Log. of $\dfrac{5}{7} = $ log. of $5 - $ log. of $7$.

(387.) *If the log. of a number be multiplied by* n, *the product is the log. of that number raised to the* $n^{th}$ *power.*

Let $d$ be the number whose log. is $x$, or $a^x = d$; then $a^{nx} = d^n$; that is, $nx$ is the log. of $d^n$.

Exs. Log. of $\overline{13}\,|^5 = 5 \times$ log. $13$. Log. $b^z = z \times$ log. $b$.

(388.) *If the log. of a number be divided by* n, *the quotient is the log. of the* $n^{th}$ *root of that number.*

Let $a^x = d$, then $a^{\frac{x}{n}} = d^{\frac{1}{n}}$, or $\dfrac{x}{n}$ is the log. of $d^{\frac{1}{n}}$.

Ex. Log. of $5^{\frac{1}{4}} = \dfrac{1}{4} \times$ log. of $5$.

(389.) The utility of a table of logarithms in arithmetical calculations will from hence be manifest; the multiplication and division of numbers being performed by the addition and subtraction of these artificial representatives; and the involution or evolution of numbers, by multiplying or dividing their logarithms by the indices of the powers or roots required.

Ex.

Ex. Let the value of $\sqrt[5]{7\sqrt{2}\times\sqrt[3]{3}}$ be required.

Log. of $7 = .845098$

$\frac{1}{2}$ log. of $2 = .150515$

$\frac{1}{3}$ log. of $3 = .1590404$

$5)\overline{1.1546534}$ sum

$.2309306 = $ log. of $1.70188$ &c. the value required.

## ON INTEREST AND ANNUITIES.

(390.) *Interest* is the consideration paid for the use or forbearance of the payment of money. The *rate* of interest is the consideration paid for the use of a certain sum for a certain time, as of £1. for one year.

When the interest of the *principal* alone, or sum lent, is taken, it is called *simple* interest; but if the interest, as soon as it becomes due, be considered as principal, and interest be charged upon the whole, it is called *compound* interest.

The *amount* is the whole sum due at the end of any time, interest and principal together.

*Discount* is the abatement made for the payment of money before it becomes due.

(391.) *To find the amount of a given sum, in any time, at simple interest.*

Let $P$ be the principal,

$r$, the interest of one pound for one year,

$n$, the

*n*, the time for which the interest is to be calculated,

*M*, the amount.

Then since the interest of a given sum, at a given rate, must be proportional to the time, 1 (year) : *n* (years) :: *r* : *nr* the interest of £1. for *n* years; and the interest of *P*£, must be *P* times as great, or *nrP*; therefore the amount $M = P + nrP$.

(392.) In this simple equation, any three of the quantities *P*, *n*, *r*, *M* being given, the fourth may be found; thus, $P = \dfrac{M}{1 + nr}$.

Ex. What sum must be paid down to receive £600. at the end of nine months, allowing 5 per cent. discount? Or, which is the same thing, what principal *P* will in nine months be equivalent, or amount to £600., allowing 5 per cent. interest?

In this case, $M = 600$, $n = \dfrac{3}{4} = .75$, $r = \dfrac{5}{100} = .05$;

hence, $P = \dfrac{M}{1 + nr} = \dfrac{600}{1 + .75 \times .05} = 578.313$ &c. £.

(393.) *To find the amount of an annuity, or pension left unpaid any number of years, allowing simple interest upon each sum or pension from the time it becomes due.*

Let *A* be the annuity; then at the end of the first year, *A* becomes due, and at the end of the second year, the interest of the first annuity is *rA* (Art. 391); at the end of this year, the principal becomes 2*A*, therefore the interest due at the end of the third year is 2*rA*,

$2rA$; in the same manner, the interest due at the end of the fourth year is $3rA$; &c. hence, the whole interest is $rA + 2rA + 3rA \ldots \ldots + \overline{n-1}.rA = n.\dfrac{n-1}{2} rA$ (Art. 212); and the sum of the annuities is $nA$; therefore the whole amount $M = nA + n.\dfrac{n-1}{2} rA$.

(394.) *Required the present value of an annuity to continue a certain number of years, allowing simple interest for the money.*

Let $P$ be the present value; then if $P$, and the annuity, at the same rate of interest, amount to the same sum, they are upon the whole of equal value. The amount of $P$, in $n$ years, is $P + nrP$ (Art. 391); and the amount of the annuity in the same time is $nA$ $+ n.\dfrac{n-1}{2}rA$; therefore $P + nrP = nA + n.\dfrac{n-1}{2}rA$, and

$$P = \frac{nA + n.\dfrac{n-1}{2} rA}{1 + nr}.$$

(395.) In this equation any three of the four quantities $P$, $A$, $n$, $r$ being given, the other may be found.

(396.) Cor. Let $n$ be infinite, then $P = \dfrac{nA}{2}$ an infinite quantity; therefore for a finite annuity to continue for ever, an infinite sum ought, according to this calculation, to be paid; a conclusion which shews the necessity of estimating the value of an annuity upon different principles.

(397.) To

(397.) *To find the amount of a given sum at compound interest.*

Let $R = \pounds 1$. together with it's interest for a year; then at the end of the first year, $R$ becomes the principal, or sum due; therefore

$1 : R :: R : R^2$, the amount in two years;

$1 : R :: R^2 : R^3$, the amount in three years; &c. in the same manner, $R^n$ is the amount in $n$ years; and if $P$ be the principal, the amount must be $P$ times as great, or $PR^n = M$.

(398.) Cor. 1. From this equation we have $P = \dfrac{M}{R^n}$.

Ex. What sum must be paid down to receive $\pounds 600$. at the end of three years, allowing 5 per cent. per ann. compound interest;

In this case, $R = 1.05$, $n = 3$, $M = 600$; and consequently $P = \dfrac{M}{R^n} = \dfrac{600}{\overline{1.05}|^3} = 518.302 \pounds$.

(399.) Cor. 2. If $P$, $R$ and $M$ be given to find $n$, we have log. $P + n \times$ log. $R =$ log. $M$; and $n =$

$$\dfrac{\text{log. } M - \text{log. } P}{\text{log. } R}$$

(400.) *To find the amount of an annuity in any number of years, at compound interest.*

Let $A$ be the annuity, or sum due at the end of the first year; then $1 : R :: A : RA$, it's amount at the end of the second year; therefore $A + RA$ is the sum due at the end of the second year; in the same manner, $1 : R :: \overline{1+R} \times A : \overline{R+R^2} \times A$, the amount of the

the two payments at the end of the third year; and $\overline{1 + R + R^2} \times A$ is the whole sum due at the end of the third year; in the same manner, $\overline{1 + R + R^2 \dots + R^{n-1}} \times A$ is the sum due at the end of $n$ years, that is, $\dfrac{R^n - 1}{R - 1} \times A = M$ (Art. 222).

(401.) Cor. In this equation, any three of the quantities being given, the fourth may be found.

(402.) *To find the present value of an annuity to be paid for* n *years, allowing compound interest.*

Let $P$ be the present value, $A$ the annuity; then since $PR^n$ is the amount of $P$ in $n$ years, and $\dfrac{R^n - 1}{R - 1} \times A$, the amount of $A$ in the same time; by the question, $PR^n = \dfrac{R^n - 1}{R - 1} \times A$, and $P = \dfrac{1 - \dfrac{1}{R^n}}{R - 1} \times A.$

(403.) Cor. 1. Any three of the quantities $P$, $A$, $R$, $n$ being given, the fourth may be found.

(404.) Cor. 2. If the number of years be infinite, $R^n$ is infinite, and $\dfrac{1}{R^n}$ vanishes; therefore $P = \dfrac{A}{R - 1}.$

Ex. If the annual rent of a freehold estate be £1., what is it's value, allowing 5 per cent. compound interest?

In this case, $A = 1$, $R - 1 = .05$; therefore the present value $P = \dfrac{1}{.05} = £20.$, or 20 years purchase.

(405.) Cor.

(405.) Cor. 3. The present value of an annuity, to commence at the expiration of $p$ years, and to continue $q$ years, is the difference between it's present value for $p+q$ years, and it's present value for $p$ years.

Ex. What is the present value of an annuity of £1., for 14 years, to commence at the expiration of 7 years, allowing 5 per cent. compound interest?

The present value for 21 years $= \dfrac{\overline{1.05}|^{21}-1}{\overline{1.05}|^{21}\times.05} =$ 12.82£.; and the present value for 7 years $= \dfrac{\overline{1.05}|^{7}-1}{\overline{1.05}|^{7}\times.05} = 5.79£.$; hence, the value of the annuity for 14 years after the expiration of 7, is 7.03£.; nearly.

## SCHOLIUM.

(406.) The method of determining the present value of an annuity at simple interest, given in Art. 394, has been decried by several eminent Arithmeticians, and in it's stead, a solution of the question has been proposed upon the following principle; " If the present value of each payment be determined separately, the sum of these values must be the value of the whole annuity."

Let $x$ be the value or price paid down for the annuity, $a$ the yearly payment, $n$ the number of years for which it is to be paid, $r$ the interest of £1. for one year. The present value of the first payment is $\dfrac{a}{1+r}$

p                                          (Art.

(Art. 392), the present value of the second payment, or of $a \pounds$. to be paid at the end of two years, is $\dfrac{a}{1+2r}$,

and so on; therefore $x = \dfrac{a}{1+r} + \dfrac{a}{1+2r} \cdots \dfrac{a}{1+nr}$.

(407.) These different conclusions arise from a circumstance which the Opponents seem not to have attended to. According to the former solution, no part of the interest of the price paid down is employed in paying the annuity, till the principal is exhausted.

Let the annuity be always paid out of the principal $x$ as long as it lasts, and afterwards out of the interest which has accrued; then $x$, $x-a$, $x-2a$, $x-3a$, &c. are the sums in hand, during the first, second, third, fourth, &c. years, the interest arising from which, $rx, rx-ra, rx-2ra, rx-3ra$, &c. that is, the whole interest, is $nrx - \overline{1+2+3\ldots n-1} \times ra$, or, $nrx - n.\dfrac{n-1}{2} ra$, which, together with the principal $x$, is equal to the sum of all the annuities; therefore

$$\overline{1+nr}.x - n.\dfrac{n-1}{2} ra = na, \text{ and } x = \dfrac{na + n.\dfrac{n-1}{2} ra}{1+nr}$$

(Art. 394).

According to the other calculation, part of the interest, as it arises, is employed in paying the annuity, but not the whole. Thus, the first payment is made by a part of the principal, and the interest of that part, which together amount to the annuity; and the other payments are made in the same manner; this is, in effect,

effect, allowing interest upon that part of the whole interest which is incorporated with the principal. According to either calculation, the seller has the advantage, since the whole, or a part of the interest will remain at his disposal till the last annuity is paid off.

If the whole interest, as it arises, be incorporated with the principal, and employed in paying the annuity, compound interest is, in effect, allowed upon the whole. Let $x$ be the price paid for the annuity, $n$ the number of years for which it is granted, and $R = 1\pounds.$ together with it's interest for one year. Then $x$ in one year amounts to $Rx$, out of which the annuity being paid, $Rx - a$ is the sum in hand at the end of the first year; $R^2x - Ra$ is the amount of this sum at the end of the second year, therefore $R^2x - Ra - a$ is the sum in hand at the end of the second year; in the same manner, $R^nx - R^{n-1}a - R^{n-2}a \ldots - a$ is the sum left, after paying the last annuity, which ought to be nothing; therefore $R^nx = R^{n-1}a + R^{n-2}a \ldots + a = $
$$\frac{R^na - a}{R - 1}, \text{ and } x = \frac{R^na - a}{R^n \times \overline{R-1}}. \quad \text{(Art. 402)}.$$

## ON THE SUMMATION OF SERIES.

(408.) We have before seen the method of determining the sums of quantities in arithmetical and geometrical progression, but when the terms increase or decrease according to other laws, different artifices must be used to obtain general expressions for their sums.

The

The methods chiefly adopted, and which may be considered as belonging to Algebra, are, 1. The method of subtraction. 2. The summation of recurring series, by the scale of relation. 3. The Differential method. 4. The method of Increments.

(409.) *The investigation of series whose sums are known by subtraction.*

### Ex. 1.

Let $1 + \dfrac{1}{2} + \dfrac{1}{3} + \dfrac{1}{4} + \&c.$ in inf. $= S$

then $\dfrac{1}{2} + \dfrac{1}{3} + \dfrac{1}{4} + \dfrac{1}{5} + \&c.$ in inf. $= S - 1$

by subtraction, $\dfrac{1}{1.2} + \dfrac{1}{2.3} + \dfrac{1}{3.4} + \&c.$ in inf. $= 1.$

### Ex. 2.

Let $1 - \dfrac{1}{2} + \dfrac{1}{3} - \dfrac{1}{4} + \&c.$ in inf. $= S$

then $- \dfrac{1}{2} + \dfrac{1}{3} - \dfrac{1}{4} + \dfrac{1}{5} - \&c.$ in inf. $= S - 1$

by subtraction, $\dfrac{3}{1.2} - \dfrac{5}{2.3} + \dfrac{7}{3.4} - \dfrac{9}{4.5} + \&c.$ in inf. $= 1.$

### Ex. 3.

Let $1 + \dfrac{1}{2} + \dfrac{1}{3} + \dfrac{1}{4} + \&c.$ in inf. $= S$

then $\dfrac{1}{3} + \dfrac{1}{4} + \dfrac{1}{5} + \dfrac{1}{6} + \&c.$ in inf. $= S - \dfrac{3}{2}$

by subt. $\dfrac{2}{1.3} + \dfrac{2}{2.4} + \dfrac{2}{3.5} + \dfrac{2}{4.6} + \&c.$ in inf. $= \dfrac{3}{2}$

or $\dfrac{1}{1.3} + \dfrac{1}{2.4} + \dfrac{1}{3.5} + \dfrac{1}{4.6} + \&c.$ in inf. $= \dfrac{3}{4}$

In

In the same manner, if $1 - \frac{1}{2} + \frac{1}{3} - \frac{1}{4} + $ &c. in inf.

$= S$, we obtain $\frac{1}{1.3} - \frac{1}{2.3} + \frac{1}{3.5} - $ &c. in inf. $= \frac{1}{4}$.

### Ex. 4.

Let $\frac{1}{1.2} + \frac{1}{2.3} + \frac{1}{3.4} + $ &c. in inf. $= S$

then $\frac{1}{2.3} + \frac{1}{3.4} + \frac{1}{4.5} + $ &c. in inf. $= S - \frac{1}{2}$

by subt. $\frac{2}{1.2.3} + \frac{2}{2.3.4} + \frac{2}{3.4.5} + $ &c. in inf. $= \frac{1}{2}$

and $\frac{1}{1.2.3} + \frac{1}{2.3.4} + \frac{1}{3.4.5} + $ &c. in inf. $= \frac{1}{4}$.

### Ex. 5.

Let $\frac{1}{m} + \frac{1}{m+r} + \frac{1}{m+2r} + \dots \frac{1}{m+n-1.r} + \frac{1}{m+nr} = S$,

then $\frac{1}{m+r} + \frac{1}{m+2r} + \frac{1}{m+3r} + \dots \frac{1}{m+nr} = S - \frac{1}{m}$

by subt. $\frac{r}{m.m+r} + \frac{r}{m+r.m+2r} + $&c. (to $n$ terms) $+ \frac{1}{m+nr} = \frac{1}{m}$

hence, $\frac{r}{m.m+r} + \frac{r}{m+r.m+2r} + $&c. (to $n$ terms) $= \frac{1}{m} - \frac{1}{m+nr}$

and $\frac{1}{m.m+r} + \frac{1}{m+r.m+2r} + $&c. (to $n$ terms) $= \frac{1}{mr} - \frac{1}{mr+nr^2}$.

If $n$ be increased without limit, $\frac{1}{mr+nr^2}$ vanishes,

and the sum of the series is $\frac{1}{mr}$.

If $m = r = 1$, we have $\frac{1}{1.2} + \frac{1}{2.3} + \frac{1}{3.4} + $ &c. (to $n$

terms) $= 1 - \frac{1}{1+n} = \frac{n}{1+n}$.

In

In the same manner,

if $\dfrac{1}{m} - \dfrac{1}{m+r} + \dfrac{1}{m+2r}\ldots \mp \dfrac{1}{m+n-1.r} \pm \dfrac{1}{m+nr} = S$

$\dfrac{2m+r}{m.m+r} - \dfrac{2m+3r}{m+r.m+2r} + \&c.$ to $n$ terms $= \dfrac{1}{m} \mp \dfrac{1}{m+nr}$.

### Ex. 6.

Let $\dfrac{1}{m.m+r} + \dfrac{1}{m+r.m+2r} + \&c.$ in inf. $= S$

then $\dfrac{1}{m+r.m+2r} + \dfrac{1}{m+2r.m+3r} + \&c. = S - \dfrac{1}{m.m+r}$

by subt. $\dfrac{2r}{m.m+r.m+2r} + \dfrac{2r}{m+r.m+2r.m+3r} + \&c. = \dfrac{1}{m.m+r}$

or $\dfrac{1}{m.m+r.m+2r} + \dfrac{1}{m+r.m+2r.m+3r} + \&c. = \dfrac{1}{2rm.m+r}$.

The sum of $n$ terms of the series, determined as in

the last example, is $\dfrac{1}{2rm.m+r} - \dfrac{1}{2r.m+nr.m+n+1.r}$.

Let $m=r=1$; then $\dfrac{1}{1.2.3} + \dfrac{1}{2.3.4} + \dfrac{1}{3.4.5} + \&c.$

in infinitum $= \dfrac{1}{4}$; and $\dfrac{1}{1.2.3} + \dfrac{1}{2.3.4} + \&c.$ to $n$ terms,

$= \dfrac{1}{4} - \dfrac{1}{2.n+1.n+2}$.

### Ex. 7.

(410.) To find the sum of the series $\dfrac{1}{2.4.6} + \dfrac{1}{4.6.8} +$

$\dfrac{1}{6.8.10} + \&c.$ in infinitum.

*When the sum of a series of this kind is required,*
*take away the last factor out of each denominator,*
*and assume the resulting series equal to* S; *and then*
*proceed as in the former examples.*

Let

Let $\dfrac{1}{2.4}+\dfrac{1}{4.6}+\dfrac{1}{6.8}+$ &c. in inf. $=S$

then $\dfrac{1}{4.6}+\dfrac{1}{6.8}+\dfrac{1}{8.10}+$ &c. $=S-\dfrac{1}{8}$

by subt. $\dfrac{4}{2.4.6}+\dfrac{4}{4.6.8}+\dfrac{4}{6.8.10}+$ &c. $=\dfrac{1}{8}$

and $\dfrac{1}{2.4.6}+\dfrac{1}{4.6.8}+\dfrac{1}{6.8.10}+$ &c. in inf. $=\dfrac{1}{32}$.

### Ex. 8.

(411.) To find the sum of the series $\dfrac{1}{m}+\dfrac{n}{m.\overline{m+a}}+$

$\dfrac{n.\overline{n+a}}{m.\overline{m+a}.\overline{m+2a}}+$ &c. in infinitum, when $n$ is less than $m$.

Let $1+\dfrac{n}{m}+\dfrac{n.\overline{n+a}}{m.\overline{m+a}}+\dfrac{n.\overline{n+a}.\overline{n+2a}}{m.\overline{m+a}.\overline{m+2a}}+$ &c. in inf. $=S$

$\dfrac{n}{m}+\dfrac{n.\overline{n+a}}{m.\overline{m+a}}+\dfrac{n.\overline{n+a}.\overline{n+2a}}{m.\overline{m+a}.\overline{m+2a}}+$ &c. $=S-1$

by subt. $\dfrac{m-n}{m}+\dfrac{n.\overline{m-n}}{m.\overline{m+a}}+\dfrac{n.\overline{n+a}.\overline{m-n}}{m.\overline{m+a}.\overline{m+2a}}+$ &c. $=1$

and $\dfrac{1}{m}+\dfrac{n}{m.\overline{m+a}}+\dfrac{n.\overline{n+a}}{m.\overline{m+a}.\overline{m+2a}}+$ &c. $=\dfrac{1}{m-n}$.

If $n=2,\ a=1$; then, $\dfrac{1}{m}+\dfrac{2}{m.\overline{m+1}.}+\dfrac{2.3}{m.\overline{m+1}.\overline{m+2}}$

$+$ &c. $=\dfrac{1}{m-2}$.

### Ex. 9.

To find the sum of the series $\dfrac{a+b}{m.\overline{m+r}.\overline{m+2r}}+$

$\dfrac{a+2b}{\overline{m+r}.\overline{m+2r}.\overline{m+3r}}+\dfrac{a+3b}{\overline{m+2r}.\overline{m+3r}.\overline{m+4r}}+$ &c. in infinitum.

Let

Let $\dfrac{x+y}{m.m+r} + \dfrac{x+2y}{m+r.m+2r} + \dfrac{x+3y}{m+2r.m+3r}$ +&c. in inf. $= S$

then $\dfrac{x+2y}{m+r.m+2r} + \dfrac{x+3y}{m+2r.m+3r} + \dfrac{x+4y}{m+3r.m+4r}$ +&c. $= S - \dfrac{x+y}{m.m+r}$

by subtr. $\dfrac{2rx-my+2ry}{m.m+r.m+2r} + \dfrac{2rx-my+3ry}{m+r.m+2r.m+3r} + \dfrac{2rx-my+4ry}{m+2r.m+3r.m+4r}$ +&c. $= \dfrac{x+y}{m.m+r}$*.

In order that this series may coincide with the proposed series, let $2rx-my+2ry = a+b$; $2rx-my+3ry = a+2b$; then $y = \dfrac{b}{r}$, and $x = \dfrac{ar+m-r.b}{2r^2}$; consequently, the sum

required, or $\dfrac{x+y}{m.m+r}$, is $\dfrac{ar+\overline{m+r}.b}{m.m+r.2r^2}$.

* Here we take for granted that the terms of the assumed series converge to 0.

Let

Let $a=3$, $b=m=r=1$; then $\dfrac{4}{1.2.3}+\dfrac{5}{2.3.4}+$

$\dfrac{6}{3.4.5}+$ &c. in inf. $=\dfrac{3+2}{1.2.2}=\dfrac{5}{4}$.

Let $a=0$, $b=1$, $m=1$, $r=2$; then $\dfrac{1}{1.3.5}+\dfrac{2}{3.5.7}$

$+\dfrac{3}{5.7.9}+$ &c. in inf. $=\dfrac{1}{8}$.

See *Philosophical Transactions*, Vol. lxxii. page 389.

(412.) Similar to the method of subtraction is the following, given by De Moivre, *Miscel. Anal.* p. 130.

*Assume a series, whose terms converge to* 0, *involving the powers of an indeterminate quantity* x; *call the sum of the series S, and multiply both sides of the equation by a binomial, trinomial, &c. which involves the powers of* x *and invariable coefficients; then if* x *be so assumed that the binomial, trinomial, &c. may vanish, and some of the first terms be transposed, the sum of the remaining series is equal to the terms so transposed.*

### Ex. 1.

Let $1+\dfrac{x}{2}+\dfrac{x^2}{3}+\dfrac{x^3}{4}+$ &c. in inf. $=S$.

Multi-

Multiplying both sides by $x-1$, we have

$$\left.\begin{array}{l} x+\dfrac{x^2}{2}+\dfrac{x^3}{3}+\dfrac{x^4}{4}+\&\text{c.} \\[2mm] -1-\dfrac{x}{2}-\dfrac{x^2}{3}-\dfrac{x^3}{4}-\dfrac{x^4}{5}-\&\text{c.} \end{array}\right\} = \overline{x-1}.S$$

or $-1+\dfrac{x}{1.2}+\dfrac{x^2}{2.3}+\dfrac{x^3}{3.4}+\&\text{c.}=\overline{x-1}.S$; and if $x=1$,

then, $-1+\dfrac{1}{1.2}+\dfrac{1}{2.3}+\dfrac{1}{3.4}+\&\text{c.}=0$; or, $\dfrac{1}{1.2}+\dfrac{1}{2.3}+$

$\dfrac{1}{3.4}+\&\text{c.}$ in inf.$=1$.

### Ex. 2.

Assume $1+\dfrac{x}{2}+\dfrac{x^2}{3}+\dfrac{x^3}{4}+\&\text{c.}$ in inf.$=S$, and multiply both sides by $x^2-1$; then

$$\left.\begin{array}{l} x^2+\dfrac{x^3}{2}+\dfrac{x^4}{3}+\&\text{c.} \\[2mm] -1-\dfrac{x}{2}-\dfrac{x^2}{3}-\dfrac{x^3}{4}-\dfrac{x^4}{5}-\&\text{c.} \end{array}\right\} = \overline{x^2-1} \times S,$$

or, $-1-\dfrac{x}{2}+\dfrac{2x^2}{1.3}+\dfrac{2x^3}{2.4}+\dfrac{2x^4}{3.5}+\&\text{c.}=\overline{x^2-1}\times S.$

Let $x^2-1=0$, then $x=1$, or $-1$; if $x=1$, then

$-1-\dfrac{1}{2}+\dfrac{2}{1.3}+\dfrac{2}{2.4}+\dfrac{2}{3.5}+\&\text{c.}$ in inf.$=0$,

that is, $\dfrac{2}{1.3}+\dfrac{2}{2.4}+\dfrac{2}{3.5}+\&\text{c.}$ in inf.$=\dfrac{3}{2}$,

and $\dfrac{1}{1.3}+\dfrac{1}{2.4}+\dfrac{1}{3.5}+\&\text{c.}$ in inf.$=\dfrac{3}{4}$.

If

If $x = -1$, then

$$-1 + \frac{1}{2} + \frac{2}{1.3} - \frac{2}{2.4} + \frac{2}{3.5} - \&c. \text{ in inf.} = 0,$$

and $\dfrac{1}{1.3} - \dfrac{1}{2.4} + \dfrac{1}{3.5} - \&c.$ in inf. $= \dfrac{1}{4}$.

### Ex. 3.

Let $1 + \dfrac{x}{2} + \dfrac{x^2}{3} + \dfrac{x^3}{4} + \&c.$ in inf. $= S$, and multiply both sides by $2x - 1$; then

$$-1 + \frac{3x}{1.2} + \frac{4x^2}{2.3} + \frac{5x^3}{3.4} + \&c. = \overline{2x-1}.S; \text{ let } 2x - 1$$

$= 0$, or $x = \dfrac{1}{2}$, and

$$-1 + \frac{3}{1.2.2} + \frac{4}{2.3.2^2} + \frac{5}{3.4.2^3} + \&c. = 0,$$

or $\dfrac{3}{1.2.2} + \dfrac{4}{2.3.2^2} + \dfrac{5}{3.4.2^3} + \&c.$ in inf. $= 1$.

(413.) If both sides of the equation be multiplied by a binomial, each term of the series obtained will have two factors in it's denominator; if by a trinomial, each term will have three factors in it's denominator; &c.

### Ex. 4.

Let $1 + \dfrac{x}{2} + \dfrac{x^2}{3} + \dfrac{x^3}{4} + \&c.$ in inf. $= S$.

multiply both sides by $\overline{2x-1}.\overline{x-1}$, or $2x^2 - 3x + 1$: then,

$$\left. \begin{array}{l} 2x^2 + \dfrac{2x^3}{2} + \dfrac{2x^4}{3} + \&c. \\[2mm] -3x - \dfrac{3x^2}{2} - \dfrac{3x^3}{3} - \dfrac{3x^4}{4} - \&c. \\[2mm] +1 + \dfrac{x}{2} + \dfrac{x^2}{3} + \dfrac{x^3}{4} + \dfrac{x^4}{5} + \&c. \end{array} \right\} = \overline{2x^2 - 3x + 1}.S$$

or

or $1 - \dfrac{5x}{2} + \dfrac{5x^2}{1.2.3} + \dfrac{6x^3}{2.3.4} + \dfrac{7x^4}{3.4.5} + \&\text{c} = \overline{2x^2 - 3x + 1}.S.$

If $x = 1$, then $1 - \dfrac{5}{2} + \dfrac{5}{1.2.3} + \dfrac{6}{2.3.4} + \&\text{c}. = 0$, and

$\dfrac{5}{1.2.3} + \dfrac{6}{2.3.4} + \dfrac{7}{3.4.5} + \&\text{c. in inf.} = \dfrac{3}{2}.$    If $x = \dfrac{1}{2}$

then $1 - \dfrac{5}{4} + \dfrac{5}{1.2.3.2^2} + \dfrac{6}{2.3.4.2^3} + \dfrac{7}{3.4.5.2^4} + \&\text{c}. = 0,$

or $\dfrac{5}{1.2.3.2^3} + \dfrac{6}{2.3.4.2^3} + \dfrac{7}{3.4.5.2^4} + \&\text{c. in inf.} = \dfrac{1}{4}.$

Ex. 5.

Let $\dfrac{1}{m} + \dfrac{x}{m+r} + \dfrac{x^2}{m+2r} + \&\text{c. in inf.} = S$;

multiply both sides by the binomial $ax - b$; then

$$\left. \begin{array}{l} \dfrac{ax}{m} + \dfrac{ax^2}{m+r} + \dfrac{ax^3}{m+2r} + \&\text{c.} \\[2mm] -\dfrac{b}{m} - \dfrac{bx}{m+r} - \dfrac{bx^2}{m+2r} - \dfrac{bx^3}{m+3r} - \&\text{c.} \end{array} \right\} = \overline{ax - b}.S,$$

or $-\dfrac{b}{m} + \dfrac{\overline{m+r}.a - mb}{m.m+r} \times x + \dfrac{\overline{m+2r}.a - \overline{m+r}.b}{m+r.m+2r} \times x^2 +$

$\&\text{c.} = \overline{ax - b}.S.$

Let $ax - b = 0$, and transpose $-\dfrac{b}{m}$; then

$\dfrac{\overline{m+r}.a - mb}{m.m+r} \times x + \dfrac{\overline{m+2r}.a - \overline{m+r}.b}{m+r.m+2r} \times x^2 + \&\text{c. in inf.}$

$= \dfrac{b}{m}.$ If the terms of this series can be made to coincide with the terms of a proposed series, the sum of the
latter

latter may be found. Thus, let the sum of the infinite

series $\dfrac{2}{1.3} \times \dfrac{1}{3} + \dfrac{3}{3.5} \times \dfrac{1}{3^2} + \dfrac{4}{5.7} \times \dfrac{1}{3^3} +$ &c. be required.

In this case, $x = \dfrac{1}{3}$, therefore $\dfrac{a}{3} - b = 0$, or $a = 3b$;

also, $m = 1$, $r = 2$, and $\overline{m+r}.a - mb = 2$, that is, $3a - b$,

or $9b - b = 2$; hence, $b = \dfrac{1}{4}$, $a = \dfrac{3}{4}$; and if these values

be substituted in the succeeding terms, the general series coincides with the proposed one, whose sum is

therefore $\dfrac{b}{m}$, or $\dfrac{1}{4}$.

To find the sum of $n$ terms of the series,

let $\dfrac{1}{m} + \dfrac{x}{m+r} + \dfrac{x^2}{m+2r} + \dots \dfrac{x^{n-1}}{\overline{m+n-1}.r} + \dfrac{x^n}{m+nr} = S*$ ;

and proceeding as before,

$$\left.\begin{array}{l} \dfrac{ax}{m} + \dfrac{ax^2}{m+r} + \dots \dfrac{ax^n}{\overline{m+n-1}.r} + \dfrac{ax^{n+1}}{m+nr} \\[2ex] -\dfrac{b}{m} - \dfrac{bx}{m+r} - \dfrac{bx^2}{m+2r} - \dots \dfrac{bx^n}{m+nr} \end{array}\right\} = \overline{ax - b}.S,$$

or, $-\dfrac{b}{m} + \dfrac{\overline{m+r}.a - mb}{m.m+r} \times x + \dfrac{\overline{m+2r}.a - \overline{m+r}.b}{m+r.m+2r} \times x^2 +$

$\dots \dfrac{ax^{n+1}}{m+nr} = \overline{ax - b}.S$ ; therefore, $\dfrac{\overline{m+r}.a - mb}{m.m+r} \times x +$

$\dfrac{\overline{m+2r}.a - \overline{m+r}.b}{m+r.m+2r} \times x^2 +$ &c. (to $n$ terms) $= \overline{ax - b}.S$

$$+ \dfrac{b}{m}$$

---

* This assumption renders the restriction in De Moivre's rule, respecting the convergency of the series, unnecessary.

$+\dfrac{b}{m}-\dfrac{ax^{n+1}}{m+nr}$. If $m=1$, $r=2$, $x=\dfrac{1}{3}$, $a=\dfrac{3}{4}$, $b=\dfrac{1}{4}$,

the series becomes $\dfrac{2}{1.3}\times\dfrac{1}{3}+\dfrac{3}{3.5}\times\dfrac{1}{3^2}+\dfrac{4}{5.7}\times\dfrac{1}{3^3}$

$+$ &c. the sum of which, to $n$ terms, is $\dfrac{1}{4}-$

$\dfrac{1}{4.\overline{1+2n}.3^n}$.

## Ex. 6.

To find the sum of $\dfrac{19}{1.2.3}\times\dfrac{1}{4}+\dfrac{28}{2.3.4}\times\dfrac{1}{8}+\dfrac{39}{3.4.5}$

$\times\dfrac{1}{16}+\dfrac{52}{4.5.6}\times\dfrac{1}{32}+$&c. in inf.

Because the factors in the denominators increase by 1, and begin from 1, assume $1+\dfrac{x}{2}+\dfrac{x^2}{3}+\dfrac{x^3}{4}+\dfrac{x^4}{5}+$&c. $=S$, and multiply both sides by $ax^2-bx+c$; then,

$$\left.\begin{array}{l} ax^2+\dfrac{ax^3}{2}+\dfrac{ax^4}{3}+\text{&c.} \\[2mm] -\,bx-\dfrac{bx^2}{2}-\dfrac{bx^3}{3}-\dfrac{bx^4}{4}-\text{&c.} \\[2mm] +\,c+\dfrac{cx}{2}+\dfrac{cx^2}{3}+\dfrac{cx^3}{4}+\dfrac{cx^4}{5}+\text{&c.} \end{array}\right\}=\overline{ax^2-bx+c}\times S,$$

or, $c+\dfrac{c-2b}{1.2}\times x+\dfrac{6a-3b+2c}{1.2.3}\times x^2+\dfrac{12a-8b+6c}{2.3.4}\times x^3$

$+\dfrac{20a-15b+12c}{3.4.5}\times x^4+$&c. in inf. $=\overline{ax^2-bx+c}.S$;

and

and since $ax^2 - bx + c = 0$, and one value of $x$ is $\frac{1}{2}$, because the powers of $\frac{1}{2}$ are involved in the terms of the proposed series, we have $\frac{a}{4} - \frac{b}{2} + c = 0$; also, $6a - 3b + 2c = 19$, and $12a - 8b + 6c = 28$; from which three equations it appears that $a = 6$, $b = 7$, $c = 2$; and if these values be substituted in the general series, we have, $2 - 3 + \frac{19}{1.2.3} \times \frac{1}{4} + \frac{28}{2.3.4} \times \frac{1}{8} + \frac{39}{3.4.5} \times \frac{1}{16} + \frac{52}{4.5.6} \times \frac{1}{32} + \&c. = 0$, or, $\frac{19}{1.2.3} \times \frac{1}{4} + \frac{28}{2.3.4} \times \frac{1}{8} + \frac{39}{3.4.5} \times \frac{1}{16} + \&c. = 1$.

The sum of $n$ terms of this series, determined as in the last example, is $1 - \dfrac{4+n}{n+1.n+2.2^{n+1}}$.

## ON RECURRING SERIES.

(414.) If each succeeding term of a decreasing series bear an invariable relation to a certain number of the preceding terms, the sum of the series may be found.

Let $a + bx + cx^2 + \&c.$ be the proposed series; call it's terms $A$, $B$, $C$, $D$, &c. and let $C = fxB + gx^2A$, $D = fxC + gx^2B$, &c. where $f + g$ is denominated the *scale of relation*; then, by the supposition,

$$A = A$$

$$A = A$$
$$B = B$$
$$C = fxB + gx^2A$$
$$D = fxC + gx^2B$$
$$E = fxD + gx^2C$$
&c.      &c.

and if the whole sum $A + B + C + D +$ &c. in inf. $= S$, we have,

$$S = A + B + fx \times \overline{S - A} + gx^2 \times S, \text{ or}$$
$$S - fxS - gx^2S = A + B - fxA \text{ ; therefore, } S =$$
$$\frac{A + B - fxA}{1 - fx - gx^2}.$$

In the same manner, if the scale of relation be $f + g +$ &c. to $n$ factors, the sum of the series is

$$\frac{A + B + C...(n) - fx \times \overline{A + B....(n-1)} - gx^2 \times \overline{A + ...(n-2)} \&c.}{1 - fx - gx^2 - hx^3....(n+1)}$$

### Ex. 1.

To find the sum of the infinite series $1 + 3x + 9x^2$ + &c. when $x$ is less than $\dfrac{1}{3}$.

Here $f = 3$, and the sum $= \dfrac{1}{1 - 3x}$.

### Ex. 2.

To find the sum of the infinite series $1 + 2x + 3x^2 + 4x^3 +$ &c. when $x$ is less than 1.

Here, $f = 2$, $g = -1$, and the sum $= \dfrac{1 + 2x - 2x}{1 - 2x + x^2}$

$$= \frac{1}{\overline{1 - x}|^2}.$$

If $x$ be equal to, or greater than 1, the series is infinite; yet we know that it arises from the division of 1 by $\overline{1 - x}|^2$, and the sum of $n$ terms may be determined.

The

The series after the $n$ first terms becomes $\overline{n+1}.x^n +$ $\overline{n+2}.x^{n+1} + \overline{n+3}.x^{n+2} + \&c.$ in which the scale of relation, as before, is $2-1$; and therefore the series arises from the fraction $\dfrac{\overline{n+1}.x^n + \overline{n+2}.x^{n+1} - 2.\overline{n+1}.x^{n+1}}{1 - 2x + x^2}$, or

$\dfrac{\overline{n+1}.x^n - nx^{n+1}}{\overline{1-x}|^2}$; therefore $1 + 2x + 3x^2 + \&c.$ to $n$

terms $= \dfrac{1 - \overline{n+1}.x^n + nx^{n+1}}{\overline{1-x}|^2}$

If the sign of $x$ be changed, $1 - 2x + 3x^2 - \&c.$ to

$n$ terms $= \dfrac{1 \mp \overline{n+1}.x^n \mp nx^{n+1}}{\overline{1+x}|^2}$, where the upper or

lower sign is to be used, according as $n$ is an even or odd number.

### Ex. 3.

To find the sum of $n$ terms of the series $1 + 3x + 5x^2 + 7x^3 + \&c.$

Suppose $f + g$ to be the scale of relation; then, $3f + g = 5$, and $5f + 3g = 7$; hence, $f = 2$, and $g = -1$; and, by trial, it appears that the scale of relation is properly determined; hence, $S = \dfrac{1 + 3x - 2x}{1 - 2x + x^2} = \dfrac{1 + x}{\overline{1-x}|^2}$.

After $n$ terms, the series becomes $\overline{2n+1}.x^n + \overline{2n+3}.x^{n+1} + \overline{2n+5}.x^{n+2} + \&c.$ which arises from the

fraction $\dfrac{\overline{2n+1}.x^n + \overline{2n+3}.x^{n+1} - 2.\overline{2n+1}.x^{n+1}}{1 - 2x + x^2}$; or,

$\dfrac{\overline{2n+1}.x^n - \overline{2n-1}.x^{n+1}}{\overline{1-x}|^2}$; hence, $1 + 3x + 5x^2 + 7x^3 +$

$\&c.$ to $n$ terms, $= \dfrac{1 + x - \overline{2n+1}.x^n + \overline{2n-1}.x^{n+1}}{\overline{1-x}|^2}$.

Q

Ex. 4.

## Ex. 4.

To find the sum of $1 + 2x + 3x^2 + 5x^3 + 8x^4 + \&c.$ in inf. when the series converges.

In this case the scale of relation is $1 + 1$, and consequently, the sum is $\dfrac{1 + 2x - x}{1 - x - x^2} = \dfrac{1 + x}{1 - x - x^2}$.

If $x$ become negative, $1 - 2x + 3x^2 - 5x^3 + \&c.$ in

inf. $= \dfrac{1 - x}{1 + x - x^2}$.

## Ex. 5.

To find the sum of $n$ terms of the series $\overline{n - 1}.x + \overline{n - 2}.x^2 + \overline{n - 3}.x^3 + \&c.$

In the series $\overline{n - 1}.x + \overline{n - 2}.x^2 + \overline{n - 3}.x^3 + \&c.$ the scale of relation is $2 - 1$; therefore it's sum in inf. is

$\dfrac{\overline{n - 1}.x + \overline{n - 2}.x^2 - 2.\overline{n - 1}.x^2}{\overline{1 - x}|^2}$, or $\dfrac{\overline{n - 1}.x - nx^2}{\overline{1 - x}|^2}$. After

$n$ terms, the series becomes $- x^{n + 1} - 2x^{n + 2} - \&c.$ the sum

of which is found, in the same manner, to be $\dfrac{- x^{n + 1}}{\overline{1 - x}|^2}$;

therefore $\overline{n - 1}.x + \overline{n - 2}.x^2 + \overline{n - 3}.x^3 + \&c.$ to $n$ terms,

$= \dfrac{\overline{n - 1}.x - nx^2 + x^{n + 1}}{\overline{1 - x}|^2}$.

Hence, $\dfrac{\overline{n - 1}.x}{n} + \dfrac{\overline{n - 2}.x^2}{n} + \&c.$ to $n$ terms, $=$

$\dfrac{\overline{n - 1}.x - nx^2 + x^{n + 1}}{n.\overline{1 - x}|^2}$.

## Ex. 6.

To find the sum of $n$ terms of the series $1^2 + 2^2 x + 3^2 x^2 + 4^2 x^3 + \&c.$

Let

Let the scale of relation be $f+g+h$; then,

$$9f+\ 4g+\ h=16$$
$$16f+\ 9g+4h=25$$
$$25f+16g+9h=36.$$

From these equations we obtain $f=3$, $g=-3$, $h=1$, which values, when substituted, produce the successive terms of the proposed series; therefore $S=$

$$\frac{1+4x+9x^2-3x-12x^2+3x^2}{1-3x+3x^2-x^3}=\frac{1+x}{\overline{1-x}\,|^3},\ \text{the sum of}$$

the series in inf. when $x$ is less than 1.

After the first $n$ terms, the series becomes $\overline{n+1}\,|^2\times x^n+$ $\overline{n+2}\,|^2\times x^{n+1}+\overline{n+3}\,|^2\times x^{n+2}+$ &c. which arises from

the fraction $\dfrac{\overline{n+1}\,|^2\times x^n+\overline{n+2}\,|^2\times x^{n+1}+\overline{n+3}\,|^2\times x^{n+2}-}{-1-3x+}$

$$\frac{3.\overline{n+1}\,|^2 x^{n+1}-3.\overline{n+2}\,|^2 x^{n+2}+3.\overline{n+1}\,|^2 x^{n+2}}{3x^2-x^3},\ \text{or,}$$

$$\frac{\overline{n+1}\,|^2.x^n-\overline{2n^2+2n-1}\times x^{n+1}+n^2x^{n+2}}{\overline{1-x}\,|^3}\ ;\ \text{and conse-}$$

quently the sum of $n$ terms of the series is

$$\frac{1+x-\overline{n+1}\,|^2 x^n+\overline{2n^2+2n-1}\times x^{n+1}-n^2x^{n+2}}{\overline{1-x}\,|^3}.$$

On this subject the Reader may consult De Moivre's *Misc. Analyt.* p. 72. And Euler's *Analys. Infinit.* C. XIII.

## ON THE DIFFERENTIAL METHOD.

(415.) In any series of quantities $a$, $b$, $c$, $d$, $e$, $f$, &c. if each term be taken from that which follows it, and the differences of these differences be taken, and so on, the following ranks of differences will be obtained;

1<sup>st</sup> Diff.

$1^{st}$ Diff. $b-a, c-b, d-c, e-d, f-e,$ &c.

$2^{d}$ Diff. $c-2b+a, d-2c+b, e-2d+c, f-2e+d,$ &c.

$3^{d}$ Diff. $d-3c+3b-a, e-3d+3c-b, f-3e+3d-c,$

$$[\&c.$$

$4^{th}$ Diff. $e-4d+6c-4b+a, f-4e+6d-4c+b,$ &c.

$5^{th}$ Diff. $f-5e+10d-10c+5b-a,$ &c.

&c. &c.

Hence it appears, that the coefficients of the quantities $a, b, c, d,$ &c. in the first term of the $n^{th}$ differences, are the coefficients of the terms of a binomial raised to the $n^{th}$ power, and that their signs are alternately positive and negative; that is, the first term of the $n^{th}$ differences is $a-nb+n.\dfrac{n-1}{2}c-n.\dfrac{n-1}{2}.\dfrac{n-2}{3}$

$d+$&c. or, $-a+nb-n.\dfrac{n-1}{2}c+n.\dfrac{n-1}{2}.\dfrac{n-2}{3}d-$&c.

according as $n$ is an even or an odd number.

Cor. If the $n^{th}$ differences vanish, the $\overline{n+1}^{th}$ term of the original series is $\pm a-nb+n.\dfrac{n-1}{2}c-$ &c. to $n$

terms; and the $\overline{n+2}^{th}$ term is $\pm b-nc+n.\dfrac{n-1}{2}d-$&c.

that is, the series recurs, and the scale of relation is

$n-n.\dfrac{n-1}{2}+n.\dfrac{n-1}{2}.\dfrac{n-2}{3}-$ &c. continued to $n$ terms.

(416.) Let $d^{I}, d^{II}, d^{III}, d^{IV},$ &c. represent the first terms in the first, second, third, fourth, &c. orders of differences; then,

$$d^{I} = b-a$$
$$d^{II} = c-2b+a$$
$$d^{III} = d-3c+3b-a$$
$$d^{IV} = e-4d+6c-4b+a$$
$$\&c.$$

and

and by transposition,

$b = a + d^\text{i}$

$c = 2b - a + d^\text{ii} = a + 2d^\text{i} + d^\text{ii}$

$d = 3c - 3b + a + d^\text{iii} = a + 3d^\text{i} + 3d^\text{ii} + d^\text{iii}$

$e = 4d - 6c + 4b - a + d^\text{iv} = a + 4d^\text{i} + 6d^\text{ii} + 4d^\text{iii} + d^\text{iv}$

&c.        &c.

from which it is manifest, that the coefficients of $a$, $d^\text{i}$, $d^\text{ii}$, $d^\text{iii}$, &c. in the expression for the $\overline{n+1}^\text{th}$ term of the series $a$, $b$, $c$, $d$, &c. are the coefficients of the terms of a binomial raised to the $n^\text{th}$ power; that is, the $\overline{n+1}^\text{th}$ term of the series is $a + nd^\text{i} + n.\dfrac{n-1}{2}d^\text{ii}$

$+ n.\dfrac{n-1}{2}.\dfrac{n-2}{3}d^\text{iii} + \text{&c.}$

(417.) Cor. 1. The $n^\text{th}$ term of the series is $a + \overline{n-1}.d^\text{i}$

$+ \overline{n-1}.\dfrac{n-2}{2}d^\text{ii} + \overline{n-1}.\dfrac{n-2}{2}.\dfrac{n-3}{3}d^\text{iii} + \text{&c.}$

### Ex.

Required the $n^\text{th}$ term of the series 1, 3, 5, &c.

$$1, \quad 3, \quad 5, \quad 7$$
$$2, \quad 2, \quad 2$$
$$0, \quad 0$$

Here, $a = 1$, $d^\text{i} = 2$, $d^\text{ii} = 0$; therefore the $n^\text{th}$ term is $1 + \overline{n-1}.2 = 2n - 1$.

(418.) Cor. 2. If the differences at length vanish, the $n^\text{th}$ term of the series will be exactly determined; but if the differences do not vanish, we can only approximate to it; and the less the differences become, when compared with the former differences, and with $n$, the nearer will the approximation be to the true value of that term.

(419.) Let

(419.) Let the proposed series be $o, a, a+b, a+b+c,$ $a+b+c+d$, &c. then,

        1$^{st}$ Diff. $a, b, c, d$, &c.

        2$^{d}$ Diff. $b-a, c-b, d-c$, &c.

        3$^{d}$ Diff. $c-2b+a, d-2c+b$, &c.

        4$^{th}$ Diff. $d-3c+3b-a$, &c.

        &c.

Let $b-a=d^{1}, c-2b+a=d^{11}, d-3c+3b-a=d^{111}$, &c. then the $\overline{n+1}^{th}$ term of the series, that is, $a+$ $b+c+d+$&c. to $n$ terms, is $na+n.\dfrac{n-1}{2}d^{1}+n.\dfrac{n-1}{2}.$ $\dfrac{n-2}{3}d^{11}+$&c. (Art. 416).

*If therefore* a, b, c, d, *&c. be the terms of any series, whose first, second, third, &c. differences are represented by* d', d'', d''', *&c. the sum of* n *terms of this series is*

$$na+n.\frac{n-1}{2}d^{1}+n.\frac{n-1}{2}.\frac{n-2}{3}d^{11}+\&c.$$

### Ex. 1.

Required the sum of the series $1+3+5+7+$&c. continued to $n$ terms.

        1,   3,   5,   7

          2,   2,   2

            0,   0

In this case, $a=1, d^{1}=2, d^{11}=0$; hence, the sum is $n+n.\overline{n-1}=n^{2}.$

### Ex. 2.

Required the sum of the series $1^{2}+2^{2}+3^{2}+4^{2}+$ &c. continued to $n$ terms.

        1,   4,   9,   16

          3,   5,   7

           2,   2

            0

                                 Here,

Here, $a=1$, $d'=3$, $d''=2$, $d'''=0$; therefore the

$$\text{sum} = n+n.\frac{n-1}{2}\times 3+n.\frac{n-1}{2}.\frac{n-2}{3}\times 2 = \frac{n.\overline{2n^2+3n+1}}{1.2.3}$$

$$= \frac{n.\overline{n+1}.\overline{2n+1}}{1.2.3}.$$

### Ex. 3.

To find the sum of the series $1^3+2^3+3^3+$&c. to $n$ terms.

$$
\begin{array}{ccccc}
1, & 8, & 27, & 64, & 125 \\
7, & 19, & 37, & 61 \\
12, & 18, & 24 \\
6, & 6 \\
0
\end{array}
$$

Here, $a=1$, $d'=7$, $d''=12$, $d'''=6$, $d^{iv}=0$; therefore the sum $=n+n.\dfrac{n-1}{2}\times 7+n.\dfrac{n-1}{2}.\dfrac{n-2}{3}\times 12 +$

$$n.\frac{n-1}{2}.\frac{n-2}{3}.\frac{n-3}{4}\times 6 = \frac{n^4+2n^3+n^2}{4}=n.\overline{\frac{n+1}{2}}\Big|^2.$$

Sometimes the sum may be more readily obtained by beginning the series with one or more cyphers; thus, to find the sum of $n$ terms of the series $1^3+2^3+3^3+$&c. take $n+1$ terms of the series $0+1^3+2^3+3^3+$&c.

$$
\begin{array}{ccccc}
0, & 1, & 8, & 27, & 64 \\
1, & 7, & 19, & 37 \\
6, & 12, & 18 \\
6, & 6 \\
0
\end{array}
$$

Here, $a=0$, $d'=1$, $d''=6$, $d'''=6$; and the sum
of

of $n+1$ terms is $\overline{n+1}.\dfrac{n}{2} + \dfrac{\overline{n+1}.n.\overline{n-1}}{1.2.3} \times 6 +$

$\dfrac{\overline{n+1}.n.\overline{n-1}.\overline{n-2}}{1.2.3.4} \times 6 = \dfrac{\overline{n+1}.n.2 + \overline{n+1}.n.\overline{n-1}.4 +}{4}$

$\dfrac{\overline{n+1}.n.\overline{n-1}.\overline{n-2}}{} = \dfrac{\overline{n+1}.n.2 + 4n - 4 + n^2 - 3n + 2}{4} =$

$\dfrac{\overline{n+1}.n.\overline{n^2+n}}{4} = \overline{\dfrac{n.\overline{n+1}}{2}}\Big|^{2}.$

The differential method may also be applied to the interpolation of series, and the quadrature of curves. See·Emerson's Tract on this subject.

# ON THE METHOD OF INCREMENTS.

(420.) Any variable quantity is called an *integral*.

The magnitude by which it is increased at one step, is called the *increment*. Thus, $1 + 2 + 3\dots\dots + m = m.\dfrac{m+1}{2}$, and the magnitude by which it increases at one step is $m+1$, which is called the increment of the integral $m.\dfrac{m+1}{2}$.

When the quantity decreases, the increment becomes negative.

(421.) *If two quantities begin to increase together, and their corresponding increments be always in the same ratio, their integrals, or the whole quantities generated, will be in that ratio.*

Let

Let the corresponding increments be $A$, $B$, $C$, &c. and $a$, $b$, $c$, &c. and let $A : a :: B : b :: C : c$ &c. $:: m : n$; then $A + B + C +$ &c. $: a + b + c +$ &c. $:: m : n$, (Art. 183).

Cor. When $m = n$, or the increments are equal, the integrals are also equal.

(422.) *If an integral be represented by the product of quantities in arithmetical progression, as* $\overline{m}.\overline{m+r}.\overline{m+2r}.\overline{m+3r}....\overline{m+n-1}.r$, *where* r *is constant, and* m *is increased at every step by* r, *the increment of this integral is* $nr \times \overline{m+r}.\overline{m+2r}.\overline{m+3r} ....\overline{m+n-1}.r$.

The first value of the integral is $m.\overline{m+r}.\overline{m+2r}.$ $\overline{m+3r}....\overline{m+n-1}.r$, and the succeeding value is $\overline{m+r}.\overline{m+2r}.\overline{m+3r}.\overline{m+4r}......\overline{m+nr}$, the difference of these values, or the increment of the integral, is $nr \times \overline{m+r}.\overline{m+2r}.\overline{m+3r}....\overline{m+n-1}.r$.

(423.) Cor. 1. Since an invariable quantity $C$ has no increment, the increment of $m.\overline{m+r}.\overline{m+2r}.\overline{m+3r}$ $....\overline{m+n-1}.r \pm C$ is also $nr \times \overline{m+r}.\overline{m+2r}.\overline{m+3r}$ $....\overline{m+n-1}.r$.

(424.) Cor. 2. Hence, if the increment of an integral be represented by $nr \times \overline{m+r}.\overline{m+2r}.\overline{m+3r}....$ $\overline{m+n-1}.r$, the integral is $m.\overline{m+r}.\overline{m+2r}.\overline{m+3r}....$ $\overline{m+n-1}.r \pm C$; where the invariable quantity $C$ must be determined by the nature of the question.

(425.) Cor.

(425.) Cor. 3. If the increment be $A \times \overline{m+r}.\overline{m+2r}.$
$\overline{m+3r}\dots \overline{m+n-1}.r$, where $A$ is invariable, the

integral is $\dfrac{A}{nr} \times \overline{m}.\overline{m+r}.\overline{m+2r}\dots \overline{m+n-1}.r \pm C.$

(Art. 421.)

(426.) If $A$ be the constant increment, and $m$ the number of times it has been taken, the integral is $mA \pm C.$

(427.) *To find, therefore, the integral of any increment, let the increment be reduced to the products of arithmetical progressionals whose common difference is the quantity by which the variable magnitude is increased at every step, and the integral of each increment will be found by multiplying it by the preceding term in the progression, and dividing it by the number of terms, thus increased, and by the common difference.*

(428.) The constant quantity which is to be added to, or subtracted from this result, in order to obtain the correct integral, must be determined by the nature of the question ; thus, when $x$, the integral obtained by the rule, is $a$, suppose the true integral is known to be $b$; then since $x + C$ is in all cases the integral, $a + C = b$, or $C = b - a$ ; therefore the correct integral is $x + b - a$.

### Ex. 1.

To find the sum of the series $1 + 2 + 3 + \&c.$ continued to $n$ terms.

The $n^{\text{th}}$ term is $n$, and the increment of the sum is $n + 1$, whose integral, according to the rule, is $\dfrac{n.\overline{n+1}}{2}.$

And

And this is the correct integral, because when $n = 1$, the sum is $\dfrac{1.2}{2} = 1$, as it ought to be.

Ex. 2.

To find the $n^{th}$ term of the series 5, 9, 16, 26, 39, &c.

Take the differences, as in Art. 415,

$$5, \quad 9, \quad 16, \quad 26, \quad 39,$$
$$4, \quad 7, \quad 10, \quad 13$$
$$3, \quad 3, \quad 3$$

It is manifest that the $n^{th}$ term of any order of differences is the increment of the $n^{th}$ term of the preceding order; therefore 3 is the increment of the $n^{th}$ term of the first differences, and $3n + C$ is the $n^{th}$ term; when $n = 1$, $3 + C = 4$, or $C = 1$; hence the $n^{th}$ term of the first differences, or the increment of the $n^{th}$ term of the original series, is $3n + 1$; consequently 'the $n^{th}$ term required is $\dfrac{3.\overline{n-1}.n}{2} + n + C$;

let $n = 1$, and $1 + C = 5$, or $C = 4$; therefore the $n^{th}$ term is $\dfrac{3.\overline{n-1}.n}{2} + n + 4$.

Ex. 3.

To find the sum of the series $1^2 + 2^2 + 3^2 + $ &c. continued to $n$ terms.

The increment of this sum is $\overline{n+1}|^2 = n.\overline{n+1} + \overline{n+1}$, and the integral, or sum, is $\dfrac{\overline{n-1}.n.\overline{n+1}}{3} + \dfrac{n.\overline{n+1}}{2} = \dfrac{\overline{2n+1}.n.\overline{n+1}}{2.3}$; which needs no correction.

Ex. 4.

To find the sum of the series $1^2 + 3^2 + 5^2 + $ &c. continued to $n$ terms.

The

The increment of the sum is $\overline{2n+1}\vert^2 = 4n^2 + 4n + 1$

$= 4n.\overline{n+1} + 1$, whose integral is $\dfrac{4.\overline{n-1}.n.\overline{n+1}}{3} + n$;

which needs no correction.

### Ex. 5.

To find the sum of $n$ terms of the series $1^3 + 2^3 + 3^3$ $+$ &c.

The increment of the sum is $\overline{n+1}\vert^3 = n^3 + 3n^2 + 3n$ $+1$, $= n.\overline{n+1}.\overline{n+2} + \overline{n+1}$, whose integral is

$\dfrac{\overline{n-1}.n.\overline{n+1}.\overline{n+2}}{4} + \dfrac{n.\overline{n+1}}{2} = \overline{\dfrac{n.\overline{n+1}}{2}}\Big\vert^2$, the sum re-

quired; which needs no correction.

The following table of *figurate* numbers is formed by making the $n^{th}$ term of each succeeding rank equal to the sum of $n$ terms of the preceding.

| | | | | | | |
|---|---|---|---|---|---|---|
| 1ˢᵗ Order | 1, | 1, | 1, | 1, | 1, | 1 |
| 2ᵈ | | 1, | 2, | 3, | 4, | 5, 6 |
| 3ᵈ | | 1, | 3, | 6, | 10, | 15, 21 |
| 4ᵗʰ | | 1, | 4, | 10, | 20, | 35, 56 |
| 5ᵗʰ | | 1, | 5, | 15, | 35, | 70,126 |
| &c. | | &c. | | | | |

### Ex. 6.

To find the sum of $n$ terms of the $m^{th}$ order of figurate numbers.

The sum of $n$ terms of the 1ˢᵗ order is $n$; therefore $n+1$ is the increment of the sum of $n$ terms of the 2ᵈ order, and it's integral, $n.\dfrac{n+1}{2}$, is the $n^{th}$ term of the

3ᵈ order; consequently $\dfrac{\overline{n+1}.\overline{n+2}}{2}$ is the $\overline{n+1}^{th}$ term,

or increment of the sum of $n$ terms, of the 3ᵈ order,

and

and it's integral, $\dfrac{\overline{n.n+1}.\overline{n+2}}{2.3}$, is the sum of $n$ terms of the $3^d$ order, or the $n^{th}$ term of the $4^{th}$ order; &c.

Thus it appears that $\dfrac{\overline{n.n+1}....\overline{n+m-1}}{1.2......m}$ is the sum which was to be found; the integrals requiring no correction.

### Ex. 7.

To find the sum of $n$ terms of the series $1.2+2.3+3.4+$ &c.

The $\overline{n+1}^{th}$ term, or increment of the sum, is $\overline{n+1}.$ $\overline{n+2}$, and consequently the integral, or sum, is $\dfrac{\overline{n.n+1}.\overline{n+2}}{3}$; which needs no correction.

### Ex. 8.

To find the sum of the $n$ first terms of a series whose $n^{th}$ term is $an^3+bn^2+cn+d$; $a, b, c, d,$ being given quantities.

Assume $A.\overline{n.n+1}.\overline{n+2}+B.\overline{n.n+1}+C.n+D=a.\overline{n+1}|^3$ $+b.\overline{n+1}|^2+c.\overline{n+1}+d$; or

$$\left.\begin{array}{l} An^3+3An^2+2An+D \\ \quad+\ Bn^2+\ Bn \\ \quad+\ Cn \end{array}\right\} = \left\{\begin{array}{l} an^3+3an^2+3an+a \\ \quad+\ bn^2+2bn+b \\ \quad+\ cn+c \\ \quad+d \end{array}\right.$$

and by equating the coefficients, $A=a$; $3A+B=3a+b$, or $3a+B=3a+b$, that is, $B=b$; $2A+B+C=3a+2b+c$, or $2a+b+C=3a+2b+c$, hence, $C=a+b+c$; also, $D=a+b+c+d$; therefore the increment of the sum is $an.\overline{n+1}.\overline{n+2}+b.n.\overline{n+1}+\overline{a+b+c}.n$ $+a+b+c+d$, and the integral $\dfrac{a.\overline{n-1}.n.\overline{n+1}.\overline{n+2}}{4}$

$+$

$$+ \frac{b.\overline{n-1}.n.\overline{n+1}}{3} + \frac{\overline{a+b+c}.\overline{n-1}.n}{2} + \overline{a+b+c+d}.n;$$

which requires no correction.

(429.) *Though in general it is convenient to reduce an increment to the products of arithmetical progressionals, in order to obtain it's integral; yet if a quantity of any other form can be found, whose increment coincides with the increment proposed, this quantity, when properly corrected, is the integral* (Art. 421).

### Ex. 1.

To find the sum of $n$ terms of the series $5+6+7+$ &c.

Let $An + Bn^2 + Cn^3 +$ &c. be the sum required; it's increment is $A.\overline{n+1} + B.\overline{n+1}|^2 + C.\overline{n+1}|^3 +$ &c. $- An - Bn^2 - Cn^3 -$ &c. and the increment of the sum is also $n+5$; therefore $A + 2Bn + B + 3Cn^2 + 3Cn + C +$ &c. $=n+5$; and by equating the coefficients, $C=0$; $2B=1$, or $B=\frac{1}{2}$; $A+B=5$, or $A=\frac{9}{2}$; hence the sum required is $\frac{9n+n^2}{2}$; which needs no correction.

### Ex. 2.

To find the number of shot in a pyramidal pile upon a square base whose side is known.

Let $n$ be the number in one side of the base; then $n^2$ is the number contained in the first square; also, since one shot in the next square, will lie between every two in the former, $n-1$ is the number contained in the side

side of the second square, and $\overline{n-1}\,|^2$ the number in that square, &c. therefore the number of shot is $n^2 +$ $\overline{n-1}\,|^2 + \overline{n-2}\,|^2 \ldots + 2^2 + 1^2$; or $1^2 + 2^2 + 3^2 \ldots + n^2$.

Suppose $An + Bn^2 + Cn^3$ to be the sum of the series; it's increment is $A.\overline{n+1} + B.\overline{n+1}\,|^2 + C.\overline{n+1}\,|^3 - An$ $- Bn^2 - Cn^3$; and the increment of the series is also $\overline{n+1}\,|^2$; therefore

$$\left.\begin{array}{r} 3\,Cn^2 + 3\,Cn + C \\ + 2\,Bn + B \\ + A \end{array}\right\} = n^2 + 2\,n + 1\ ;$$

and by equating the coefficients, $3\,C = 1$, or $C = \dfrac{1}{3}$, $3\,C + 2\,B = 2$, or $B = \dfrac{1}{2}$; $C + B + A = 1$, or $\dfrac{1}{3} + \dfrac{1}{2} + A$ $= 1$, hence $A = \dfrac{1}{6}$; and the sum of the series is $\dfrac{n}{6} + \dfrac{n^2}{2}$ $+ \dfrac{n^3}{3}$; which requires no correction. If $An + Bn^2 +$ $Cn^3 + Dn^4 + $&c. be assumed for the sum of the series, it is evident, from the process, that $D$, and the coefficient of every succeeding term, vanishes.

(430.) *If* $\dfrac{1}{m.m+r....m+n-1.r}$ *represent an in-* *tegral, where* r *is constant, and* m *increases at every* *step by* r, *it's increment is* $\dfrac{-nr}{m.m+r....m+nr}$.

For, the next value of the integral is

$\dfrac{1}{m+r.m+2r....m+nr}$, from which if the preceding

value

value be taken, the difference or increment, is

$$\frac{m - \overline{m + nr}}{m.\overline{m + r}....\overline{m + nr}} = \frac{-nr}{m.\overline{m + r}....\overline{m + nr}}.$$

(431.) Cor. 1. Hence, if $\dfrac{-nr}{m.\overline{m + r}....\overline{m + nr}}$ represent

an increment, it's integral is $\dfrac{1}{m.\overline{m + r}....\overline{m + n - 1.r}}$.

(432.) Cor. 2. If the increment be $\dfrac{A}{m.\overline{m + r}....\overline{m + nr}}$

the integral is $\dfrac{-A}{nr} \times \dfrac{1}{m.\overline{m + r}...\overline{m + n - 1.r}}$ (Art. 421).

That is, if an increment can be reduced to the form

$\dfrac{A}{m.\overline{m + r}....\overline{m + nr}}$, it's integral may be found by re-

jecting the last term in the denominator, and dividing by the number of terms left, and by the common difference taken negatively.

This integral must be corrected as in Art. 428.

### Ex. 1.

To find the sum of $m$ terms of the series $\dfrac{1}{1.2.3} +$

$\dfrac{1}{2.3.4} + \dfrac{1}{3.4.5} \div$ &c.

The $\overline{m + 1}^{\text{th}}$ term, or increment of the sum, is

$\dfrac{1}{\overline{m + 1}.\overline{m + 2}.\overline{m + 3}}$, whose integral is $\dfrac{-1}{\overline{2.m + 1}.\overline{m + 2}} + C$;

and when $m = 1$, the integral is $\dfrac{1}{1.2.3}$; therefore $\dfrac{-1}{2.2.3}$

$+ C =$

$$+ C = \frac{1}{1.2.3}, \text{ and } C = \frac{1}{1.2.3} + \frac{1}{2.2.3} = \frac{3}{12} = \frac{1}{4};$$

hence, the correct integral, or sum required, is $\frac{1}{4}$ –

$$\frac{1}{2.\overline{m+1}.\overline{m+2}}.$$

### Ex. 2.

To find the sum of $n$ terms of the series $\dfrac{5}{1.2.3.4}$ +

$\dfrac{7}{2.3.4.5} + \dfrac{9}{3.4.5.6}$ + &c.

The $\overline{n+1}^{\text{th}}$ term, or increment of the sum, is

$$\frac{2n+5}{\overline{n+1}.\overline{n+2}.\overline{n+3}.\overline{n+4}} = \frac{2.\overline{n+1}+3}{\overline{n+1}.\overline{n+2}.\overline{n+3}.\overline{n+4}} =$$

$$\frac{2}{\overline{n+2}.\overline{n+3}.\overline{n+4}} + \frac{3}{\overline{n+1}.\overline{n+2}.\overline{n+3}.\overline{n+4}}; \text{ and the}$$

integral or sum $= \dfrac{-1}{\overline{n+2}.\overline{n+3}} - \dfrac{1}{\overline{n+1}.\overline{n+2}.\overline{n+3}} + C;$

and when $n = 1$, the sum is $\dfrac{5}{1.2.3.4}$; therefore

$-\dfrac{1}{3.4} - \dfrac{1}{2.3.4} + C = \dfrac{5}{1.2.3.4};$ hence, $C = \dfrac{5+2+1}{1.2.3.4} = \dfrac{1}{3},$

and the correct integral, or sum required, is $\dfrac{1}{3}$ –

$$\frac{1}{\overline{n+2}.\overline{n+3}} - \frac{1}{\overline{n+1}.\overline{n+2}.\overline{n+3}} = \frac{1}{3} - \frac{1}{\overline{n+1}.\overline{n+3}}.$$

R　　　　　　　　　　Ex. 3.

## Ex. 3.

To find the sum of $n$ terms of the series $\dfrac{1}{1.3} + \dfrac{1}{2.4}$

$+ \dfrac{1}{3.5} + \dfrac{1}{4.6} + \&c.$

The increment of the sum is $\dfrac{1}{\overline{n+1}.\overline{n+3}} =$

$\dfrac{n+2}{\overline{n+1}.\overline{n+2}.\overline{n+3}} = \dfrac{\overline{n+1}+1}{\overline{n+1}.\overline{n+2}.\overline{n+3}} = \dfrac{1}{\overline{n+2}.\overline{n+3}} +$

$\dfrac{1}{\overline{n+1}.\overline{n+2}.\overline{n+3}}$ ; therefore the sum is $- \dfrac{1}{\overline{n+2}} -$

$\dfrac{1}{2.\overline{n+1}.\overline{n+2}} + C$; let $n = 1$, then $- \dfrac{1}{3} - \dfrac{1}{2.2.3} + C =$

$\dfrac{1}{1.3}$ ; hence, $C = \dfrac{2}{3} + \dfrac{1}{12} = \dfrac{9}{12} = \dfrac{3}{4}$, and the sum

required is $\dfrac{3}{4} - \dfrac{1}{n+2} - \dfrac{1}{2.\overline{n+1}.\overline{n+2}}$.

## Ex. 4.

To find the sum of $n$ terms of the series $\dfrac{1}{1.3} +$

$\dfrac{1}{3.5} + \dfrac{1}{5.7} + \&c.$

The increment of the sum is $\dfrac{1}{\overline{2n+1}.\overline{2n+3}}$ ; assume

$2v = 2n + 1$, then $v$ increases as fast as $n$, i.e. it increases

by 1 at every step; and $\dfrac{1}{\overline{2n+1}.\overline{2n+3}} = \dfrac{1}{2v.\overline{2v+2}}$

$= \dfrac{1}{4}$

$$= \frac{1}{4} \times \frac{1}{v.v+1}, \text{ whose integral is } -\frac{1}{4v} + C = -\frac{1}{4n+2}$$

$+ C$; and when $n = 1$, we have $-\frac{1}{6} + C = \frac{1}{3}$, or $C = \frac{1}{2}$ ;

therefore the sum required is $\frac{1}{2} - \frac{1}{4n+2}$.

They who wish to prosecute this subject farther, may consult Dr. Waring's *Fluxions*, Sterling's *Summation of Series*, and Emerson's *Method of Increments*.

## ON CHANCES.

(433.) *If an event may take place in* n *different ways, and each of these be equally likely to happen, the* probability *that it will take place in a specified way is* properly represented by $\frac{1}{n}$, *certainty being represented by unity : Or, which is the same thing, if the value of certainty be unity, the* value of the expectation *that the event will happen in a specified way is* $\frac{1}{n}$.

For, the sum of all the probabilities is certainty, or unity, because the event must take place in some one of the ways, and the probabilities are equal; therefore each of them is $\frac{1}{n}$.

(434.) Cor. If the value of certainty be $a$, the value of the expectation is $\frac{a}{n}$. But in the following articles we suppose the value of certainty to be unity.

R 2                 (435.) *If*

(435.) *If an event may happen in* a *ways, and fail in* b *ways, any of these being equally probable, the chance of it's happening is* $\dfrac{a}{a+b}$, *and the chance of it's failing is* $\dfrac{b}{a+b}$.

The chance of it's happening must, from the nature of the supposition, be to the chance of it's failing, as $a$ to $b$; therefore the chance of it's happening : chance of it's happening, together with the chance of it's failing :: $a : a+b$; and the event must either happen or fail, consequently the chance of it's happening together with the chance of it's failing is certainty; hence, the chance of it's happening : certainty :: $a : a+b$; and the chance of it's happening

$$= \frac{a}{a+b}.$$

Also, since the chance of it's happening together with the chance of it's failing is certainty, which is represented by unity, $1 - \dfrac{a}{a+b}$, that is, $\dfrac{b}{a+b}$ is the chance of it's failing.

### Ex. 1.

(436.) The probability of throwing an ace with a single die, in one trial, is $\dfrac{1}{6}$; the probability of not throwing an ace is $\dfrac{5}{6}$; the probability of throwing either an ace or a duce, is $\dfrac{2}{6}$; &c.

Ex. 2.

## Ex. 2.

(437.) If $n$ balls, $a$, $b$, $c$, $d$, &c. be thrown promiscuously into a bag, and a person draw out one of them, the probability that it will be $a$ is $\frac{1}{n}$; the probability that it will be either $a$ or $b$ is $\frac{2}{n}$.

## Ex. 3.

(438.) The same supposition being made, if two balls be drawn out, the probability that these will be $a$ and $b$ is $\frac{2}{n.n-1}$.

For there are $n.\frac{n-1}{2}$ combinations of $n$ things taken two and two together (Art. 230); and each of these is equally likely to be taken; therefore the probability that $a$ and $b$ will be taken is $\frac{1}{n.\frac{n-1}{2}}$, or $\frac{2}{n.n-1}$.

## Ex. 4.

(439.) If 6 white and 5 black balls be thrown promiscuously into a bag, and a person draw out one of them, the probability that this will be a white ball is $\frac{6}{11}$; and the probability that it will be a black ball is $\frac{5}{11}$.

(440.) From the Bills of Mortality in different places, tables have been constructed which shew how

how many persons, upon an average, out of a certain number born, are left at the end of each year, to the extremity of life. From such tables, the probability of the continuance of a life, of any proposed age, is known.

### Ex. 1.

(441.) To find the probability that an individual of a given age will live one year.

Let $A$ be the number, in the tables, of the given age, $B$ the number left at the end of the year; then $\dfrac{B}{A}$ is the probability that the individual will live one year; and $\dfrac{A-B}{A}$ the probability that he will die in that time (Art. 435). In Dr. Halley's Tables, out of 586 of the age of 22, 579 arrive at the age of 23; hence, the probability that an individual aged 22 will live one year is $\dfrac{579}{586}$ or $\dfrac{83}{84}$ nearly; and $\dfrac{7}{586}$, or $\dfrac{1}{84}$ nearly, is the probability that he will die in that time.

### Ex. 2.

(442.) To find the probability that an individual of a given age will live any number of years.

Let $A$ be the number, in the tables, of the given age, $B, C, D, \ldots X$, the number left at the end of $1, 2, 3, \ldots t$, years; then $\dfrac{B}{A}$ is the probability that

the

the individual will live one year; $\dfrac{C}{A}$ the probability

that he will live 2 years; and $\dfrac{X}{A}$ the probability that he

will live $t$ years. Also, $\dfrac{A-B}{A}$, $\dfrac{A-C}{A}$, $\dfrac{A-X}{A}$, are the

probabilities that he will die in 1, 2, $t$, years.

These conclusions follow immediately from Art. 435.

(443.) *If two events be independent of each other,
and the probability that one will happen be* $\dfrac{1}{m}$*, and
the probability that the other will happen be* $\dfrac{1}{n}$*,
the probability that they will both happen is* $\dfrac{1}{mn}$*.*

For, each of the $m$ ways in which the first can happen
or fail, may be combined with each of the $n$ ways
in which the other can happen or fail, and thus form
$mn$ combinations, and there is only one in which
both can happen; therefore the probability that this
will be the case is $\dfrac{1}{mn}$ (Art. 433.)

(444.) Cor. 1. The probability that both do not
happen is $1-\dfrac{1}{mn}$, or $\dfrac{mn-1}{mn}$. For, the probability
that they both happen, together with the probability
that they do not both happen, is certainty; therefore
if from unity, the probability that they both happen
be

be subtracted, the remainder is the probability that
they do not both happen.

(445.) Cor. 2. The probability that they will both
fail is $\dfrac{\overline{m-1}.\overline{n-1}}{mn}$. For, the probability that the first

will fail is $\dfrac{m-1}{m}$, and the probability that the second

will fail is $\dfrac{n-1}{n}$; therefore the probability that they

will both fail is $\dfrac{m-1}{m} \times \dfrac{n-1}{n}$, or $\dfrac{\overline{m-1}.\overline{n-1}}{mn}$.

(446.) Cor. 3. The probability that one will happen
and the other fail is $\dfrac{m+n-2}{mn}$. For, the probability

that the first will happen and the second fail is $\dfrac{1}{m} \times$

$\dfrac{n-1}{n}$, and the probability that the first will fail and

the second happen is $\dfrac{m-1}{m} \times \dfrac{1}{n}$, and the sum of these,

or $\dfrac{m+n-2}{mn}$, is the probability that one will happen

and the other fail.

(447.) Cor. 4. If there be any number of independent events, and the probabilities of their happening
be $\dfrac{1}{m}, \dfrac{1}{n}, \dfrac{1}{r}$, &c. respectively, the probability that they

will all happen is $\dfrac{1}{mnr \text{ &c.}}$. For, the probability that

the

the two first will happen is $\dfrac{1}{mn}$, and the probability

that the two first and third will happen is $\dfrac{1}{mnr}$ ; and

the same proof may be extended to any number of

events. When $m = n = r$ &c. the probability is $\dfrac{1}{m^v}$, $v$

being the number of events.

## Ex. 1.

(448.) Required the probability of throwing an ace
and then a deuce with one die.

The chance of throwing an ace is $\dfrac{1}{6}$, and the chance

of throwing a deuce in the second trial is $\dfrac{1}{6}$; therefore

the chance of both happening is $\dfrac{1}{36}$.

## Ex. 2.

(449.) If 6 white and 5 black balls be thrown promiscuously into a bag, what is the probability that
a person will draw out first a white, and then a
black ball?

The probability of drawing a white ball first is $\dfrac{6}{11}$

(Art. 439), and this being done, the probability of

drawing a black ball, is $\dfrac{5}{10}$, or $\dfrac{1}{2}$, because there are 5

white and 5 black balls left; therefore the probability

required is $\dfrac{6}{11} \times \dfrac{1}{2} = \dfrac{3}{11}$. Or we may reason thus;

unless

unless the person draw a white ball first, the whole is at an end ; therefore the probability that he will have a chance of drawing a black ball is $\frac{6}{11}$, and when he has this chance, the probability of it's succeeding is $\frac{5}{10}$, or $\frac{1}{2}$; therefore the probability that both these events will take place is $\frac{6}{11} \times \frac{1}{2}$, or $\frac{3}{11}$.

## Ex. 3.

(450.) The same supposition being made, what is the chance of drawing a white ball and then two black balls?

The probability of drawing a white ball and then a black one is $\frac{3}{11}$ (Art. 449); when these two are removed, there are 5 white and 4 black balls left; and the probability of drawing a black ball, out of these, is $\frac{4}{9}$; therefore the probability required is $\frac{3}{11} \times \frac{4}{9} = \frac{4}{33}$.

## Ex. 4.

(451.) Required the probability of throwing an ace, with a single die, in two trials.

The chance of failing the first time is $\frac{5}{6}$, and the chance of failing the next is $\frac{5}{6}$; therefore the chance

of

of failing twice together is $\frac{25}{36}$, and the chance of not failing, both times, is $1 - \frac{25}{36}$, or $\frac{11}{36}$.

## Ex. 5.

(452.) In how many trials may a person undertake, for an even wager, to throw an ace with a single die?

Let $x$ be the number of trials; then, as in the last Art. the chance of failing $x$ times together is $\overline{\frac{5}{6}}\Big|^x$, and this, by the question, is equal to the chance of happening, or $\overline{\frac{5}{6}}\Big|^x = \frac{1}{2}$; hence, $x \times \log. \frac{5}{6} = \log. \frac{1}{2}$, or $x \times \overline{\log. 5 - \log. 6} = \log. 1 - \log. 2$, and $x = \frac{\log. 1 - \log. 2}{\log. 5 - \log. 6}$ $= \frac{\log. 2}{\log. 6 - \log. 5}$, since $\log. 1 = 0$; i. e. $x = 3.8$, nearly.

## Ex. 6.

(453.) To find the probability that two individuals, $P$ and $Q$, whose ages are known, will live a year.

Let the probability that $P$ will live a year, determined by Art. 441, be $\frac{1}{m}$; and the probability that $Q$ will live a year, $\frac{1}{n}$; then the probability that they will both be alive at the end of that time is $\frac{1}{m} \times \frac{1}{n}$ or $\frac{1}{mn}$.

Ex. 7.

# Ex. 7.

(454.) To find the probability that one of them, at least, will be alive at the end of any number of years.

The probability that $P$ will die in a year is $\dfrac{m-1}{m}$, and the probability that $Q$ will die is $\dfrac{n-1}{n}$; therefore the probability that they will both die is $\dfrac{\overline{m-1}.\overline{n-1}}{mn}$, and the probability that they will not both die is $1 - \dfrac{\overline{m-1}.\overline{n-1}}{mn}$, or $\dfrac{m+n-1}{mn}$.

In the same manner, if $\dfrac{1}{p}$ be the probability that $P$ will live $t$ years, and $\dfrac{1}{q}$ the probability that $Q$ will live the same time (Art. 442); the probability that one of them, at least, will be alive at the end of the time is $1 - \dfrac{\overline{p-1}.\overline{q-1}}{pq}$, or $\dfrac{p+q-1}{pq}$.

(455.) If the probability of an event's happening in one trial be represented by $\dfrac{a}{a+b}$ (Art. 435), to find the probability of it's happening once, twice, three times, &c. exactly, in $n$ trials.

The probability of it's happening in any one particular trial being $\dfrac{a}{a+b}$, the probability of it's failing in all the other $n-1$ trials is $\dfrac{b^{n-1}}{\overline{a+b}\,|^{n-1}}$ (Arts. 435. 447);

447); therefore the probability of it's happening in one particular trial, and failing in the rest, is $\dfrac{a\,b^{n-1}}{\overline{a+b}|^{n}}$; and since there are $n$ trials, the probability that it will happen in some one of these, and fail in the rest, is $n$ times as great, or $\dfrac{n\,a\,b^{n-1}}{\overline{a+b}|^{n}}$. The probability of it's happening in any two particular trials, and failing in all the rest, is $\dfrac{a^{2}b^{n-2}}{\overline{a+b}|^{n}}$, and there are $n.\dfrac{n-1}{2}$ ways in which it may happen twice in $n$ trials and fail in all the rest (Art. 230); therefore the probability that it will happen twice in $n$ trials is $\dfrac{n.\dfrac{n-1}{2}\,a^{2}b^{n-2}}{\overline{a+b}|^{n}}$. In the same manner, the probability of it's happening exactly three times is $\dfrac{n.\dfrac{n-1}{2}.\dfrac{n-2}{3}a^{3}b^{n-3}}{\overline{a+b}|^{n}}$; and the probability of it's happening exactly $t$ times is

$$\dfrac{n.\dfrac{n-1}{2}.\dfrac{n-2}{3}\ldots\ldots\dfrac{n-t+1}{t}a^{t}b^{n-t}}{\overline{a+b}|^{n}}.$$

(456.) Cor. 1. The probability of the event's failing exactly $t$ times in $n$ trials may be shewn, in the same way, to be

$$\dfrac{n.\dfrac{n-1}{2}.\dfrac{n-2}{3}\ldots\ldots\dfrac{n-t+1}{t}a^{n-t}b^{t}}{\overline{a+b}|^{n}}.$$

(457.) Cor. 2. The probability of the event's happening

pening *at least t* times in *n* trials is

$$\frac{a^n + na^{n-1}b + n.\dfrac{n-1}{2}a^{n-2}b^2 \dots \text{to } n - t + 1 \text{ terms}}{\overline{a+b}|^n}.$$

For, if it happen every time, or fail only once, twice, .... $n-t$ times, it happens $t$ times; therefore the whole probability of it's happening at least $t$ times, is the sum of the probabilities of it's happening every time, of failing only once, twice, .... $n-t$ times; and the sum of these probabilities is

$$\frac{a^n + na^{n-1}b + n.\dfrac{n-1}{2}\dot{a}^{n-2}b^2 \dots \text{to } n - t + 1 \text{ terms}}{\overline{a+b}|^n}.$$

## Ex. 1.

(458.) What is the probability of throwing an ace, twice, at least, in three trials, with a single die?

In this case, $n = 3$, $t = 2$, $a = 1$, $b = 5$; and the probability required is $\dfrac{1 + 3.5}{6.6.6} = \dfrac{16}{216} = \dfrac{2}{27}$.

## Ex. 2.

(459.) What is the probability that out of five individuals, of a given age, three, at least, will die in a given time?

Let $\dfrac{1}{m}$ be the probability that any one of them will die in the given time (Art. 442); then we have given the probability of an event's happening in one instance, to find the probability of it's happening three times in five instances.

In

In this case, $a = 1$, $b = m - 1$, $n = 5$, $t = 3$ ; therefore the probability required is $\dfrac{1 + 5.\overline{m - 1} + 10.\overline{m - 1}|^2}{m^5}$.

## SCHOLIUM.

(460.) Much more might be said on a subject so extensive as the doctrine of chances ; the Learner will however find the principal grounds of calculation in Articles 433, 435, 443, 455, and 457, and if he wish for farther information, he may consult De Moivre's work on this subject. It may not be improper to caution him against applying principles which, on the first view, may appear self-evident, as there is no subject in which he will be so likely to mistake as in the calculation of probabilities. A single instance will shew the danger of forming a hasty judgement, even in the most simple case. The probability of throwing an ace with one die is $\frac{1}{6}$, and since there is an equal probability of throwing an ace in the second trial, it might be supposed that the probability of throwing an ace in two trials is $\frac{2}{6}$.

This is not a just conclusion (Art. 451); for, it would follow, by the same mode of reasoning, that in six trials a person could not fail to throw an ace. The error, which is not easily seen, arises from a tacit supposition that there must necessarily be a second trial, which is not the case if an ace be thrown in the first.

ON

## ON LIFE ANNUITIES.

(461.) *To find the present value of an annuity of*
*£1. to be continued during the life of an individual*
*of a given age, allowing compound interest for the*
*money.*

Let $r$ be the amount of £1., in one year; $A$ the
number of persons, in the tables, of the given age;
$B, C, D$, &c. the number left at the end of 1, 2, 3,
&c. years (Art. 440); then $\frac{B}{A}$ is the value of the life

for one year, $\frac{C}{A}, \frac{D}{A}$, &c. it's value for 2, 3, &c.

years; and the series must be continued to the end
of the tables. Now the present value of £1., to be paid

at the end of one year, is $\frac{1}{r}$ (Art. 398); but it is only

to be paid on condition that the annuitant is alive at the

end of the year, of which event the probability is $\frac{B}{A}$;

therefore the present value of the conditional annuity

is $\frac{B}{Ar}$ (Art. 434); in the same manner, the present

value of the second year's annuity is $\frac{C}{Ar^2}$; the present

value of the third year's annuity is $\frac{D}{Ar^3}$; &c. there-

fore the whole value required is $\frac{1}{A} \times \overline{\frac{B}{r} + \frac{C}{r^2} + \frac{D}{r^3}} + $ &c.

to the end of the tables.

(462.) De

(462.) De Moivre supposes, that out of eighty-six persons born, one dies every year, till they are all extinct.

This supposition is sufficiently exact, if our calculations be made for any age above ten, as will appear from an inspection of the tables ; and on this supposition, the sum of the series $\dfrac{1}{A} \times \dfrac{\overline{B}}{r} + \dfrac{C}{r^2} + \dfrac{D}{r^3} + $ &c. may be found.

Let $n$ be the number of years which any individual wants of 86 ; then will $n$ be the number of persons living, of that age, out of which one dies every year ; and $\dfrac{n-1}{n}, \dfrac{n-2}{n}, \dfrac{n-3}{n}$, &c. will be the probabilities of his living, 1, 2, 3, &c. years ; hence, the present value of an annuity of £1., to be paid during his life, is $\dfrac{n-1}{nr}$

$+ \dfrac{n-2}{nr^2} + \dfrac{n-3}{nr^3} + $ &c. continued to $n$ terms. The sum of the series $\dfrac{\overline{n-1}.x}{n} + \dfrac{\overline{n-2}.x^2}{n} + \dfrac{\overline{n-3}.x^3}{n}$ to $n$ terms, was found Art. 414. Ex. 3. to be $\dfrac{\overline{n-1}.x - nx^2 + x^{n+1}}{n.\overline{1-x}|^2}$ ;

let $x = \dfrac{1}{r}$, and the sum of the series, $\dfrac{n-1}{nr} + \dfrac{n-2}{nr^2} +$

$\dfrac{n-3}{nr^3} +$ &c. to $n$ terms, is $\dfrac{\overline{n-1}.r - n + \dfrac{1}{r^{n-1}}}{n.\overline{r-1}|^2}$ ; the present value of the annuity.

s      (463.) Cor. 1.

(463.) Cor. 1. This expression for the sum is the

same with $\dfrac{nr-n}{\overline{n.r-1}|^2} - \dfrac{r-\dfrac{1}{r^{n-1}}}{\overline{n.r-1}|^2}$, or $\dfrac{1}{r-1} - \dfrac{r}{n} \times \dfrac{1-\dfrac{1}{r^n}}{\overline{r-1}|^2}$; let

$P$ be the present value of an annuity of $\pounds 1$., to continue

certain for $n$ years, then $P = \dfrac{1-\dfrac{1}{r^n}}{r-1}$ (Art. 402) ; and

the expression becomes $\dfrac{1-\dfrac{r}{n}P}{r-1}$.

(464.) Cor. 2. The present value of the annuity to continue for ever, from the death of the proposed individual, is $\dfrac{rP}{n.r-1}$.

For, the whole present value of the annuity to continue for ever, is $\dfrac{1}{r-1}$ (Art. 404) ; and if from this, it's value for the life of the individual be taken, the remainder $\dfrac{rP}{n.r-1}$ is the present value of the annuity to continue for ever, from the time of his death.

(465.) *To find the present value of an annuity of $\pounds 1$., to be paid as long as two specified individuals are both living.*

Find, by Art. 453, the probability that they will both be alive at the expiration of 1, 2, 3, &c. years, to the end of the tables : call these probabilities $a, b, c$, &c.

and

and $r$ the amount of $£1$., in one year; then $\dfrac{a}{r}+\dfrac{b}{r^2}+$ $\dfrac{c}{r^3}+$ &c. is the present value of the annuity required. (See Art. 461).

(466.) *To find the present value of an annuity of $£1$., to be paid as long as either of two specified individuals is living.*

Find, by Art. 454, the probability that they will not both be extinct in 1, 2, 3, &c. years, to the end of the tables, and call these probabilities $A$, $B$, $C$, &c. then the present value of the annuity is $\dfrac{A}{r}+\dfrac{B}{r^2}$ $+\dfrac{C}{r^3}+$ &c. (See Art. 461).

(467.) COR. If the annuity be $M£$., the present value is $M$ times as great as in the former case, or $M\times$ $\overline{\dfrac{A}{r}+\dfrac{B}{r^2}+\dfrac{C}{r^3}}+$ &c.

(468.) These are the mathematical principles on which the values of annuities for lives are calculated, and the reasoning may easily be applied to every proposed case. But in practice, these calculations, as they require the combination of every year of each life with the corresponding years of every other life concerned in the question, will be found extremely laborious, and other methods must be adopted when expedition is required. Writers on this subject, are De Moivre, Maseres, Simpson, Price, Morgan, and Waring.

THE END OF PART III.

THE

# ELEMENTS OF ALGEBRA.

## PART IV.

### THE APPLICATION OF ALGEBRA TO GEOMETRY.

(469.) THE signs made use of in algebraical calculations being general, the conclusions obtained by
their assistance may, with great ease and convenience,
be transferred from abstract magnitudes to every class
of particular quantities; thus, the relation of lines,
surfaces, or solids, may generally be deduced from the
principles of Algebra, and many properties of these
quantities discovered, which could not have been
derived from principles purely geometrical.

(470.) *Simple algebraical quantities may be represented by lines.*

Any line *AB*, may be taken at pleasure to represent
one quantity *a*, but if we have a second quantity, *b*, to
represent, we must take a line which has to the
former line, the same ratio that *b* has to *a*.

Instead

Instead of saying $AB$ represents $a$, we may say $AB = a$, supposing $AB$ to contain as many linear units as $a$ contains numeral ones.

(471.) *When a series of algebraical quantities is to be represented on one line, and each of them measured from the same point, the positive quantities being represented by lines taken in one direction, the negative quantities must be represented by lines taken in the opposite direction.*

Let $a$ be the greatest of these quantities, then $a - x$ may, by the variation of $x$, become equal to each of them in succession. Let $AB$ be the given line, and $A$ the point from which the quantities are to be measured; take $AB = a$; and since $a - x$ must be

measured from $A$, $BD$ must be taken in the contrary direction $= x$, then $AD = a - x$; and that $a - x$ may successively coincide with each quantity in the series, beginning with the greatest positive quantity, $x$ must increase; therefore $BD$, which is equal to $x$, must increase; and when $x$ is greater than $a$, $BD$ is greater than $AB$, and $AD$, which represents the negative quantity $a - x$, lies in the opposite direction from $A$.

(472.) COR. 1. If the algebraical value of a line be found to be negative, the line must be measured in a direction opposite to that which, in the investigation, we supposed to be positive.

(473.) COR. 2. If quantities be measured upon a line from it's intersection with another, the positive quantities being taken in one direction, the negative quantities must be taken in the other.

(474.) If

(474.) If a fourth proportional, to lines representing $p$, $q$, $r$ be taken, it will represent $\dfrac{qr}{p}$; and if $p = 1$, it will represent $qr$; if also, $q$ and $r$ be equal, it will represent $q^2$.

(475.) If a mean proportional between lines representing $a$ and $b$ be taken, it will represent $\sqrt{ab}$, which, when $a = 1$, becomes $\sqrt{b}$. Hence it appears, that any possible algebraical quantities may be represented by lines; and conversely, lines may be expressed algebraically; and if the relations of the algebraical quantities be known, the relations of the lines are known.

(476.) *The relations of surfaces to each other may be expressed algebraically.*

Let the sides $AB$, $AC$ of the rectangle $AD$ contain the linear units $a$, $b$, respectively; then $ab$ will be the number of superficial units contained in the

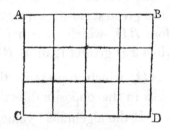

area. For, every unit in $AB$, or $a$, has $b$ units in the area, corresponding to it; consequently there are, upon the whole, $ab$ units in the area. Thus $ab$ is a proper representation of the rectangle $AD$; and by reducing other surfaces, to rectangles, their algebraical values may be found.

Cor.

Cor. Hence, the product of the two quantities $a$ and $b$, is often called their *rectangle;* and when $b$ is equal to $a$, this product is called the *square* of $a$.

(477.) In the same manner, if $a$, $b$, $c$ represent the linear units in the three sides of a rectangular parallelepiped, $abc$ will be the number of solid units contained in the figure; and consequently solids may be compared, by comparing their algebraical values.

(478.) If the line $PM$ move parallel to itself upon the indefinite line $AP$, and at the same time increase or decrease, the point $M$ will trace out a straight line, or a curve. $AP$ is called the *abscissa,* and $PM$ the *ordinate;* and the straight line, or curve, is said to be the *locus* of the point $M$.

The nature of the curve depends upon the relation of $AP$ to $PM$; and this relation, when expressed algebraically, is called the *equation to the curve.*

(479.) *Having given the nature, or construction of the curve, it's equation may be found.*

Let $BM$ be a *straight line* cutting $AP$ in a given

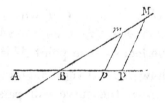

angle at $B$, the relation of $AP$ to $PM$ is expressed by a simple equation.

Suppose $AP = x$, $PM = y$, $AB = a$; then since the angles at $B$, $P$ and $M$ are invariable, $BP$ bears an invariable ratio to $PM$, let this be the ratio of $b : c$.

Then

Then since $BP = AP - AB = x - a$, we have $x - a$ : $y :: b : c$, and $by = cx - ca$, or $by - cx + ca = 0$.

(480.) Cor. A simple equation belongs to a straight line; because, by altering the values of $b$, $c$, and $a$, and taking $x$ and $a$, positive or negative, as the case requires, the equation $by - cx + ca = 0$ may be made to coincide with any proposed simple equation.

(481.) *To find the equation to the* Parabola.

Let a point $S$ be taken without the right line $CB$, and let the indefinite line $SM$ revolve about the point

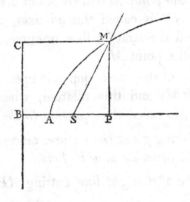

$S$ in the plane $SBC$; also, let $CM$, which is perpendicular to $CB$, cut $SM$ in $M$; then, if $SM$ be always equal to $CM$, the locus of the point $M$ is a parabola.

Through $S$ draw $BSP$ at right angles to $CB$, and if $SB$ be bisected in $A$, the curve will pass through $A$, as appears by the construction; draw $MP$ perpendicular to $BP$, and let $AP = x$, $PM = y$, $AS = a$; then $SP^2 + PM^2 = (SM^2 = CM^2 =) \ BP^2$, or $\overline{x - a}^2 + y^2$ $= \overline{x + a}^2$; that is, $x^2 - 2ax + a^2 + y^2 = x^2 + 2ax + a^2$, or $y^2 = 4ax$.

(482.) *To*

(482.) *To find the equation to the* Ellipse.

Let two indefinite lines *SM*, *HM*, revolve, in a given plane, about the points *S*, *H*, and cut each

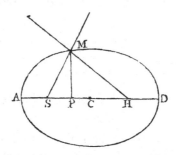

other in *M*, in such a manner, that $SM + MH$ may be an invariable quantity; then the locus of the point *M* is an ellipse.

Bisect *SH* in *C*, and from *M* draw *MP* perpendicular to *SH*, or *SH* produced; let $CP = x$, $PM = y$, $CS = c$, $SM + MH = 2a$. Then $\sqrt{SP^2 + PM^2} = SM$, and $\sqrt{HP^2 + PM^2} = HM$; therefore $\sqrt{SP^2 + PM^2} + \sqrt{HP^2 + PM^2} = SM + HM$, or $\sqrt{\overline{c - x}^2 + y^2} + \sqrt{\overline{c + x}^2 + y^2} = 2a$; hence, $\sqrt{\overline{c - x}^2 + y^2} = 2a - \sqrt{\overline{c + x}^2 + y^2}$, and squaring both sides, $c^2 - 2cx + x^2 + y^2 = 4a^2 - 4a \times \sqrt{\overline{c + x}^2 + y^2} + c^2 + 2cx + x^2 + y^2$; that is, by transposition, $4a^2 + 4cx = 4a \sqrt{\overline{c + x}^2 + y^2}$, or $a^2 + cx = a \sqrt{\overline{c + x}^2 + y^2}$; and again squaring both sides, $a^4 + 2a^2cx + c^2x^2 = a^2c^2 + 2a^2cx + a^2x^2 + a^2y^2$, or $a^2y^2 = a^4 - a^2c^2 - \overline{a^2 - c^2} \times x^2$; let $a^2 - c^2 = b^2$, then $a^2y^2 = a^2b^2 - b^2x^2$, and $y^2 = \frac{b^2}{a^2} \times \overline{a^2 - x^2}$.

(483.) Cor. 1.

(483.) Cor. 1. If $S$ and $H$ coincide, $c = 0$; hence, $a = b$, and $y^2 = a^2 - x^2$, the equation to a circle.

(484.) Cor. 2. When $x = + a$, or $- a$, then $y = 0$; therefore taking $CA = CD = a$, the curve passes through $A$ and $D$.

(485.) Cor. 3. If $AP = z$, then $x = a - z$; therefore

$$y^2 = \frac{b^2}{a^2} \times \overline{a^2 - a^2 + 2az - z^2} = \frac{b^2}{a^2} \times \overline{2az - z^2} \; ;$$ the equation which expresses the relation between $AP$ and $PM$.

(486.) Cor. 4. If $AS$ be finite, and $SM + MH$ be indefinitely increased, the limit to which the curve approaches is, at all finite distances from $S$, a *parabola.*

In this case $z^2$ vanishes when compared with $2az$; therefore the limit to which the equation approaches is $y^2 = \frac{b^2}{a^2} \times 2az$; also, $b^2 = a^2 - c^2 = \overline{a+c} \times \overline{a-c}$, and since the difference between $a$ and $c$ is finite and $a$ is infinite, $a + c$ is ultimately equal to $2a$; hence, $b^2 = 2a \times AS$; therefore $y^2 = \dfrac{2a \times AS}{a^2} \times 2az = 4AS \times z$; the equation to the parabola.

(487.) *To find the equation to the* Hyperbola.

Let two indefinite lines $SM$, $HM$ revolve, in a

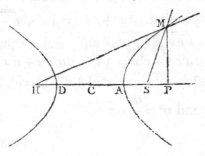

given

given plane, about the points $S$, $H$, and cut each other in $M$, in such a manner that $HM - SM$ may be a given quantity; then the locus of the point $M$ is an *Hyperbola*.

Bisect $SH$ in $C$, and draw $MP$ perpendicular to $HS$, or $HS$ produced; let $CP = x$, $PM = y$, $SC = c$, $HM - SM = 2a$. Then by proceeding as in Art. 482, $\sqrt{\overline{x + c}\,^2 + y^2} - \sqrt{\overline{x - c}\,^2 + y^2} = 2a$, and $a^2 y^2 = \overline{c^2 - a^2}.x^2$ $- a^2 c^2 - a^4$; in this case, $2c$ is greater than $2a$ (Euc. 20. 1), let therefore $b^2 = c^2 - a^2$, then $a^2 y^2 = b^2 x^2$ $- a^2 b^2$, or $y^2 = \dfrac{b^2}{a^2} \times \overline{x^2 - a^2}$.

(488.) Cor. 1. The equation to the ellipse, $y^2 = \dfrac{b^2}{a^2} \times$ $\overline{a^2 - x^2}$, becomes the equation to the Hyperbola, if $b^2$ be supposed to be negative.

(489.) Cor. 2. The equation $y^2 = \dfrac{b^2}{a^2} \times \overline{a^2 - x^2}$ may be considered as the equation to any conic section: it is the equation to an Ellipse, when $b^2$ is positive; to a Parabola, when $b^2$ is infinite (Art. 486); and to an Hyperbola, when $b^2$ is negative.

(490.) Cor. 3. If $SM - HM = 2a$, a figure, similar and equal to the former, will be traced out, which is called the opposite Hyperbola.

(491.) Cor. 4. If $x = \pm a$, then $y = 0$; therefore taking $CA = CD = a$, the curve passes through $A$ and $D$.

(492.) Cor. 5. If $AP = z$, then $CP$, or $x = z + a$, and

and $x^2 - a^2 = z^2 + 2az$; hence, $y^2 = \dfrac{b^2}{a^2} \times \overline{z^2 + 2az}$, the equation which expresses the relation between $AP$ and $PM$.

(493.) Cor. 6. In the opposite Hyperbola, $x = z - a$; therefore $x^2 - a^2 = z^2 - 2az$, and $y^2 = \dfrac{b^2}{a^2} \times \overline{z^2 - 2az}$.

(494.) *To find the equation to the* Cissoid of Diocles.

Let $AB$ be the diameter of a semicircle $ANB$; from the points $R$ and $P$, taken always at equal

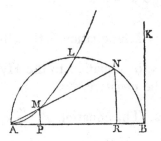

distances from $A$ and $B$, draw $RN$, $PM$, at right angles to $AB$, and join $AN$ meeting $PM$ in $M$; the point $M$ will trace out a curve called the Cissoid of Diocles.

From the nature of the circle, $AR \times RB = RN^2$ and by the construction, $AR \times RB = PB \times AP$; also, from the similar triangles $APM$, $ARN$, $AP : PM ::$ $AR : RN$, or $AP : PM :: PB : RN = \dfrac{PB \times PM}{AP}$;

and $RN^2 = \dfrac{PB^2 \times PM^2}{AP^2} = PB \times AP$; therefore $PB \times$

$$PM^2 =$$

$PM^2 = AP^3$. Let $AB = b$, $AP = x$, $PM = y$; then $\overline{b-x} \times y^2 = x^3$, the equation required.

(495.) *To find the equation to the* Conchoid of Nicomedes.

Let $AB$ be a line given in position, and about any point $C$, taken without it, let the indefinite line $CM$

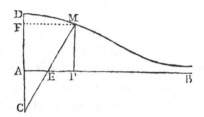

revolve, and cut $AB$ in $E$; then if $EM$ be taken always of the same length, the point $M$ will trace out a curve which is called the Conchoid of Nicomedes. Draw $CAD$ and $MP$ at right angles to $AB$, and $MF$ parallel to it; let $CA = a$, $AD = EM = b$, $AP = x$, $PM = y$. Then from the similar triangles $CFM$, $MPE$, $CF (a+y)$ : $FM (x)$ :: $MP (y)$ : $PE = \dfrac{xy}{a+y}$;

and $EM^2 = EP^2 + PM^2$, that is, $b^2 = \dfrac{x^2 y^2}{\overline{a+y}^2} + y^2$;

therefore $\overline{a+y}^2 \times b^2 = x^2 y^2 + \overline{a+y}^2 \times y^2$, or $\overline{a+y}^2 \times \overline{b^2 - y^2} = x^2 y^2$, the equation to the curve.

If $EM$ be measured in the opposite direction from $E$, the equation to the curve is $\overline{a \sim y}^2 \times \overline{b^2 - y^2} = x^2 y^2$.

(496.) *To*

(496.) *To find the equation to the* Logarithmic Curve.

If in the indefinite line *AE*, we take *AB*, *BC*, *CD*, &c. always equal to each other, and ordinates

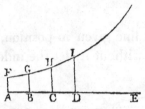

*AF, BG, CH, DI,* &c. be drawn at right angles to *AE*, and in geometrical progression, the curve *FGHI* &c. which passes through their extremities, is called the Logarithmic Curve. From the nature of logarithms (Art. 385), any abscissa *AC* is the logarithm of the corresponding ordinate *CH*, in a system which depends upon the magnitudes of *AF* and *BG*, supposing *AB* given ; in the same system, let 1 be the logarithm of *a* ; also, let $AC = x$, $CH = y$; then $x = $ log. $y$, and $1 = $ log. $a$, or $x = x \times$ log. $a$ ; therefore log. $y = x \times$ log. $a = $ log. $a^x$ ; hence, $y = a^x$, the equation to the curve.

(497.) *Having given the relation between one abscissa* CP, *and ordinate* PM, *in a curve, to find the*

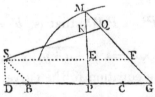

*relation between the abscissa* SQ, *which is measured from a given point* S *in a given direction, and the ordinate* QM, *which is inclined to* PM *at a given angle.*

Suppose

Suppose $PM$ perpendicular to $CP$; produce $MQ$ and $DPC$ till they meet in $G$, draw $SB$, $SD$, $SF$, respectively parallel to $MG$, $MP$, $DC$; and let sin. $\angle$ $KSE = p$; sin. $\angle SKE$, or $MKQ = m$; sin. $\angle KMQ$ $= q$; sin. $\angle MQK = n$; sin. $\angle MFE$, or $MGP = s$; to rad. 1; $SB = FG = d$, $DC = f$, $CP = x$, $PM = y$, $SQ = z$, $QM = v$.

Then in the triangle $SQF$, $s : p :: z : \dfrac{pz}{s} = QF$;

hence, $GM = GF + FQ + QM = d + \dfrac{pz}{s} + v$; and in

the triangle $MGP$, $1 : s :: d + \dfrac{pz}{s} + v : sd + pz + sv = $ $PM = y$.

Also, in the triangle $MKQ$, $m : q :: v : \dfrac{qv}{m} = KQ$, and

$SK = SQ - KQ = z - \dfrac{qv}{m}$; again, in the triangle $SKE$,

$1 : m :: z - \dfrac{qv}{m} : mz - qv = SE = DP$; hence, $f -$

$\overline{mz - qv} = f - mz + qv = CP = x$; and if these values of $x$ and $y$ be substituted in the equation which represents the relation of $CP$ to $PM$, an equation is obtained which represents the relation of $SQ$ to $QM$.

(498.) Cor. 1. Since the values of $x$ and $y$ are represented in simple terms of $z$ and $v$, the equation to the curve will rise to the same number of dimensions, whatever abscissa and ordinate are taken.

(499.) Cor. 2. From the principles of Trigonometry it appears, that $m$, $n$, and $s$ may be found in terms of $p$ and $q$; therefore in the values of $x$ and $y$, before obtained, there are only four independent invariable quantities $d$, $f$, $p$, and $q$.

(500.) Cor. 3.

(500.) Cor. 3. If the curve be a conic section whose centre is $C$ and axis $CP$, then $a^2y^2 = b^2 \times \overline{a^2 - x^2}$ (Art. 489), and substituting for $x$ and $y$ their values, we have $a^2 \times \overline{sd + pz + sv}\vert^2 = b^2 \times \overline{a^2 - f - mz + qv}\vert^2$, or arranging the terms according to the dimensions of $v$,

$$\left.\begin{array}{l} s^2a^2 \\ +q^2b^2 \end{array}\right)v^2 + \left.\begin{array}{l} 2spa^2 \\ -2mqb^2 \end{array}\right)zv + \left.\begin{array}{l} 2s^2da^2 \\ +2fqb \end{array}\right)v \\ \left.\begin{array}{l} +m^2b^2 \\ +p^2a^2 \end{array}\right)z^2 + \left.\begin{array}{l} 2spda^2 \\ -2fmb^2 \end{array}\right)z + \left.\begin{array}{l} s^2d^2a^2 \\ -b^2 \times \overline{a^2 - f^2} \end{array}\right) \Bigg\} = 0.$$

(501.) Cor. 4. The equation obtained in the last article may be made to coincide with any equation of two dimensions $Av^2 + Bzv + Cv + Dz^2 + Ez + F = 0$, by equating the coefficients of the corresponding terms; because we shall have six equations to determine the six independent quantities, $a^2$, $b^2$, $d$, $f$, $p$, $q$. Hence it follows, that every equation of two dimensions belongs to some conic section.

(502.) *Having given the equation which expresses the relation between the abscissa and ordinate, the curve may be described.*

For, any abscissa being assumed, the corresponding values of the ordinate are known from the equation; and thus, by assuming different values of the abscissa, the curve may be traced out.

(503.) Ex. 1. If $ay = bx + cd$ be the proposed equa-

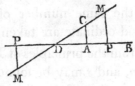

tion, it belongs to a right line (Art. 480). Let the abscissa be measured from the point $A$, along the line *AB*;

$AB$; then, when $x=0$, we have $y=\dfrac{cd}{a}$; from $A$, therefore, draw $AC$ making a finite angle with $AB$, and equal to $\dfrac{cd}{a}$, and the line which belongs to the proposed equation must pass through $C$. Also, if $y=0$, then $x=\dfrac{-cd}{b}$; take therefore, upon the line $ADP$, $AD=\dfrac{cd}{b}$, and the line to which the equation belongs must pass through $D$; therefore $DCM$ is that line.

(504.) Cor. If $AP$ be taken to represent any value of $x$, and the ordinate $PM$ be drawn parallel to $AC$, $PM$ will represent the corresponding value of $y$.

(505.) Ex. 2. Let the equation to the curve be $ax=y^2$; then, when $x=0$, we have $y=0$, or the curve passes through $A$; when $x$ is positive, $y=\pm\sqrt{ax}$, and when $x$ is infinite, these values are still possible; therefore the curve has two infinite arcs lying the same way from $A$; but when $x$ is negative, $y$

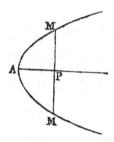

becomes impossible; therefore no part of the curve lies the other way.

T

(506.) Ex. 3.

(506.) Ex. 3. Let the equation to the curve be $xy = ab$; then, when $x$ is indefinitely small, $y$ is indefinitely great, and when $x$ is positive and indefinitely great, $y$ is positive and indefinitely small; therefore the curve will have

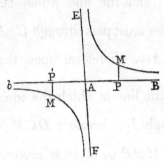

two infinite arcs between the lines $AE$ and $AB$; also, when $x$ is negative, $y$ is negative, and when infinite, $y$ is infinitely small, and when $x$ is infinitely small, $y$ is infinitely great; therefore the curve will have two infinite arcs between $Ab$ and $AF$.

These lines $EF$, $Bb$, which continually approach to the curve, and whose distances from it become, at length, less than any that can be assigned, but which produced ever so far do not meet it, are called *Asymptotes*.

(507.) Ex. 4. Let $x^4 - a^2x^2 + a^2y^2 = 0$; then, $y = \pm \frac{x}{a}\sqrt{a^2 - x^2}$; and when $x$ is nothing, $y$ is nothing, or the curve passes through $A$, the point from which $x$

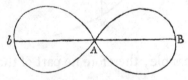

is measured. When $x = \pm a$, then $y = 0$; therefore the

the curve passes through $B$, and $b$, supposing $AB = Ab = \pm a$; but if $x$ be greater than $a$, $y$ becomes impossible; therefore no part of the curve lies beyond $B$ or $b$.

(508.) Ex. 5. To find the conic section to which any proposed quadratic equation belongs.

Let $ay^2 + \overline{b+cx}.y + d + ex + fx^2 = 0$, a general equation of two dimensions, be the proposed quadratic; then $y^2 + \dfrac{b+cx}{a} xy = -\dfrac{d+ex+fx^2}{a}$; and completing the

square, $y^2 + \dfrac{b+cx}{a} \times y + \overline{\dfrac{b+cx}{2a}}\Big|^2 = \overline{\dfrac{b+cx}{2a}}\Big|^2 - \dfrac{d+ex+fx^2}{a}$

$= \dfrac{b^2 + 2bcx + c^2x^2 - 4ad - 4aex - 4afx^2}{4a^2} =$

$\dfrac{\overline{c^2 - 4af} \times x^2 + \overline{2bc - 4ae} \times x + b^2 - 4ad}{4a^2}$ and extracting

the square root, $y + \dfrac{b+cx}{2a} =$

$\pm \dfrac{\sqrt{c^2 - 4af.x^2 + 2bc - 4ae.x + b^2 - 4ad}}{2a}$, and $y =$

$\dfrac{-\overline{b+cx} \pm \sqrt{c^2 - 4af.x^2 + 2bc - 4ae.x + b^2 - 4ad}}{2a}.$

Hence, 1. If $c^2 - 4af$ be positive, then, when $\pm x$ is infinite, $y$ has four possible values; therefore the curve has four infinite arcs, or it is the *hyperbola*.

2. If $c^2 - 4af = 0$, the curve has only two infinite arcs; because, when $\overline{2bc - 4ae}.x + b^2 - 4ad$ becomes negative, the values of $y$ are impossible; and the curve is a *parabola*. But if $2bc - 4ae$ be also $= 0$, then $y =$

$-b$

$$\frac{-\overline{b+cx} \pm \sqrt{b^2 - 4ad}}{2a}; \text{ and when } b^2 \text{ is greater than}$$

$4ad$, the curve becomes a *right line*.

3. If $c^2 - 4af$ be negative, the curve has no infinite arc; for, when $\pm x$ is infinite, the values of $y$ are impossible; hence the curve is an *ellipse*.

3. If $c^2 - 4af$ be negative, $2bc - 4ae = 0$, and $b^2 - 4ad$ be also 0, or negative, all the values of $y$ are impossible; in this case the ellipse wholly vanishes.

## ON THE CONSTRUCTION OF EQUATIONS.

(509.) The relation between the abscissa and ordinate of a conic section is expressed by a quadratic equation, in which, for every different value of the abscissa, there are two corresponding values of the ordinate, and if the abscissa be so drawn, and the conic section so constructed, that it's equation coincides with a proposed quadratic, the two ordinates will be the roots of that quadratic, which may be determined to a tolerable degree of accuracy by actual measurement.

Let $MCM$ be a circle (a figure more easily described than any other conic section) whose centre is $A$ and radius $AM$; take $AP$ an abscissa, $PM$ an ordinate meeting the circle in $M$ and $M$; join $AM$, and draw $MB$ at right angles to $AP$; let $AP = x$, $PM = y$, $AM = r$, and the cosine of the angle $APM$ (to the radius 1) $= c$; then $1 : c :: PM : PB = c \times PM = cy$, and

and $AM^2 = AP^2 + PM^2 - 2AP \times BP$, or $r^2 = x^2 + y^2 -$

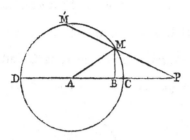

$2cxy$; that is, $y^2 - 2cxy + x^2 - r^2 = 0$; which equation may be made to coincide with any proposed quadratic.

Ex. Let the roots of the equation $y^2 - py + q = 0$ be required.

Here $2cx = p$, and $x^2 - r^2 = q$; and since there are three undetermined quantities $c$, $x$, and $r$, and only two conditions to be answered, one of these quantities may be assumed, of any finite magnitude, at pleasure: suppose $c = 1$, then $x = \dfrac{p}{2}$, $r^2 = x^2 - q = \dfrac{p^2}{4} - q$, and

$r = \sqrt{\dfrac{p^2}{4} - q}$; and since the cosine of the $\angle APM =$ radius, $PM$ coincides with $PAD$; let therefore a circle be described with the radius $\sqrt{\dfrac{p^2}{4} - q}$, cutting the line $DAP$ in $D$ and $C$; take $AP = \dfrac{p}{2}$, and the roots of the equation are $PC$ and $PD$.

(510.) The intersections of two conic sections, may be determined by a biquadratic equation, and if the figures be so drawn that this biquadratic coincides with a proposed biquadratic, the roots of the latter equation

equation may be found by measuring the ordinates which determine the points of intersection.

Let $M \overset{\prime}{M} \overset{\prime\prime}{M} \overset{\prime\prime\prime}{M}$ be a parabola whose axis is $AP$, $M \overset{\prime}{M} \overset{\prime\prime}{M} \overset{\prime\prime\prime}{M}$ a circle whose centre is $C$ and radius $CM$, cutting the parabola in the points $M$, $\overset{\prime}{M}$, $\overset{\prime\prime}{M}$, $\overset{\prime\prime\prime}{M}$;

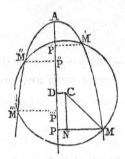

from these points draw the ordinates to the axis, $MP$, $\overset{\prime}{M}\overset{\prime}{P}$, $\overset{\prime\prime}{M}\overset{\prime\prime}{P}$, $\overset{\prime\prime\prime}{M}\overset{\prime\prime\prime}{P}$, and from $C$ draw $CD$ perpendicular to the axis, and $CN$ parallel to it, meeting $PM$ in $N$. Let $AD = a$, $DC = b$, $CM = n$, the parameter of the parabola $= p$, $AP = x$, $PM = y$; then $px = y^2$; also, $CM^2 = CN^2 + NM^2$, or $n^2 = \overline{x - a}\rvert^2 + \overline{y - b}\rvert^2$, i. e. $x^2 - 2ax + a^2 + y^2 - 2by + b^2 = n^2$; and substituting for $x$ it's value $\dfrac{y^2}{p}$, and arranging the terms according to the dimensions of $y$, we obtain $y^4 - \overline{2pa - p^2}.y^2 - 2bp^2y + p^2 \times \overline{a^2 + b^2 - n^2} = 0$, a biquadratic equation whose roots are $PM$, $\overset{\prime}{P}\overset{\prime}{M}$, $\overset{\prime\prime}{P}\overset{\prime\prime}{M}$, and $\overset{\prime\prime\prime}{P}\overset{\prime\prime\prime}{M}$, and which may be made to coincide with any proposed biquadratic whose second term is wanting.

Ex. Let the roots of the equation $y^4 - qy^2 + ry - s = 0$ be required.

Assume

Assume $p = 1$; then $2a - 1 = q$, or $a = \dfrac{q+1}{2}$; $-2b = r$,

or $b = \dfrac{-r}{2}$; $a^2 + b^2 - n^2 = -s$, or $n^2 = a^2 + b^2 + s$, and

consequently, $n = \sqrt{a^2 + b^2 + s}$; describe a parabola

whose parameter is 1, and, in the axis, take $AD = \dfrac{q+1}{2}$;

draw $DC$ at right angles to it, and $= \dfrac{-r}{2}$; from the

centre $C$, with the radius $\sqrt{a^2 + b^2 + s}$, describe the circle

$M\overset{\prime}{M}\overset{\prime\prime}{M}\overset{\prime\prime\prime}{M}$ cutting the parabola in the points $M$, $\overset{\prime}{M}$,

$\overset{\prime\prime}{M}$, $\overset{\prime\prime\prime}{M}$, then the ordinates to the axis, $PM$, $\overset{\prime}{P}\overset{\prime}{M}$, $\overset{\prime\prime}{P}\overset{\prime\prime}{M}$,

and $\overset{\prime\prime\prime}{P}\overset{\prime\prime\prime}{M}$ are the roots sought.

(511.) When $DC$ represents a negative quantity, the ordinates on the same side of the axis with $C$ represent the negative roots of the equation; and the contrary.

(512.) COR. 1. If the circle touch the parabola, two roots of the equation are equal; if it cut it only in two points, or touch it in one, two roots are impossible; and if the circle fall wholly within or without the parabola, or, if $a^2 + b^2 + s$ be negative, all the roots are impossible.

(513.) COR. 2. If $a^2 + b^2 = n^2$, or the circle pass through the point $A$, the last term of the equation $p^2 \times \overline{a^2 + b^2 - n^2} = 0$; therefore $y^4 - \overline{2pa - p^2}.y^2 - 2bp^2y = 0$, or $y^3 - \overline{2pa - p^2}.y - 2bp^2 = 0$, a cubic equation, which may be made to coincide with a proposed cubic whose second term is wanting, and the ordinates $PM$, $\overset{\prime\prime}{P}\overset{\prime\prime}{M}$, $\overset{\prime\prime\prime}{P}\overset{\prime\prime\prime}{M}$ are its roots.

(514.) COR. 3.

(514.) Cor. 3. If $a^2 + b^2 - n^2 = 0$, and also $b = 0$, the equation becomes $y^2 - \overline{2pa - p^2} = 0$; by means of which, any quadratic, wanting the second term, may be solved. In this case, the circle passes through the vertex of the parabola, and it's centre falls in the axis.

These solutions may be obtained, and nearly in the same manner, by means of any two of the conic sections.

(515.) If the roots of a cubic equation, $x^3 - qx + r = 0$, be possible, they may be found by means of a table of cosines.

Let $DAC$ be an angle whose cosine, to the radius $m$, is $x$; in $AD$, take $AB = m$; from $B$ as a centre, with the radius $BA$, describe a circle cutting $AM$ in $C$, and

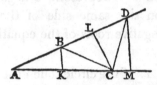

from $C$, with the same radius, describe a circle cutting $AD$ in $D$; join $BC$, $CD$, and draw $BK$, $DM$ at right angles to $AM$, and $CL$ at right angles to $AD$. Then, the triangles $BAC$ and $BCD$ being isosceles, the angles $BAC$ and $BCA$ are equal, as also $CBD$ and $CDB$; and the perpendiculars $BK$, $CL$ bisect the bases $AC$, $BD$. Also, $\angle DBC = \angle BAC + \angle BCA = 2 \angle BAC$, and $\angle DCM = \angle CAD + \angle CDA = \angle CAD + \angle CBD = \angle CAD + 2 \angle CAD = 3 \angle CAD$. Let $CM$, the cosine of the $\angle DCM$ to the radius $m$, be called $c$; then, from the similar triangles $ABK$, $ACL$, $AB : AK ::$ $AC : AL$, or $m : x :: 2x : \dfrac{2x^2}{m} = AL$, and $AL - AB$

$=$

$$= \frac{2x^2}{m} - m = BL; \text{ hence, } AD, \text{ or } AL + BL, = \frac{4x^2}{m} - m;$$

again, $AB : AK :: AD : AM$, or $m : x :: \frac{4x^2}{m} - m : \frac{4x^3}{m^2}$

$-x = AM$, and $AM - AC = CM = \frac{4x^3}{m^2} - 3x = c$, there-

fore $4x^3 - 3m^2x = m^2c$, and $4x^3 - 3m^2x - m^2c = 0$.

Let the equation $4x^3 - 3m^2x - m^2c = 0$, or $x^3 - \frac{3m^2}{4}x -$

$\frac{m^2c}{4} = 0$, be made to coincide with the equation $x^3 - qx +$

$r = 0$; that is, let $\frac{3m^2}{4} = q$, and $\frac{m^2c}{4} = -r$; or $m =$

$\sqrt{\frac{4q}{3}}$, and $c = -\frac{3r}{q}$; then, from a table of cosines,

find the angle whose cosine is $-\frac{3r}{q}$, to the radius

$\sqrt{\frac{4q}{3}}$, and the cosine of one third of this angle, to

the same radius, is one value of $x$.

(516.) Cor. 1. If $A$ be the arc whose cosine is $c$, and
$P$ the whole circumference, $c$ is also the cosine of

$A + P$, or $A + 2P$; therefore the cosines of $\frac{A+P}{3}$, and

$\frac{A+2P}{3}$ are also values of $x$.

(517.) Cor. 2. Since the radius is greater than the

cosine, $\sqrt{\frac{4q}{3}}$ is greater than $-\frac{3r}{q}$, or $\frac{4q}{3}$ is greater

than $\frac{9r^2}{q^2}$; that is, $\frac{q^3}{27}$ is greater than $\frac{r^2}{4}$; therefore this
solution can only be applied when the roots of the
cubic equation are possible. (See Art. 331.)

GENERAL

# GENERAL PROPERTIES OF CURVE LINES.

(518.) A curve is said to be of $n$ dimensions, when the equation belonging to it rises to $n$ dimensions.

Let $y^n - \overline{a\,x + b}.y^{n-1} + \overline{c\,x^2 + d\,x + e}.y^{n-2} - \&c..... + g\,x^n + h\,x^{n-1} + l\,x^{n-2} + \&c. = 0$, a general equation of $n$ dimensions, express the relation between the abscissa and ordinate of a curve, then for every different value of $x$, there are $n$ values of $y$; therefore the ordinate will cut the curve in $n$, or in $n-2$, $n-4$, &c. points, according as the equation has $n$, or $n-2$, $n-4$, &c. possible roots.

(519.) Cor. 1. Hence, if the equation be of an odd number of dimensions, the curve will have, at least, one infinite arc on each side of the point from which the abscissæ are measured; for, whatever be the value of $x$, there is, at least, one possible value of $y$ corresponding to it (Art. 278).

(520.) Cor. 2. $ax + b$ is the sum of the ordinates, $cx^2 + dx + e$ the sum of the products of any two, &c. and $gx^n + hx^{n-1} + lx^{n-2} + \&c.$ is the product of all the ordinates (Art. 271).

(521.) *If two parallel lines,* $\overset{\prime}{M}\overset{m}{M}$, $\overset{\prime}{N}\overset{m}{N}$, *be drawn*

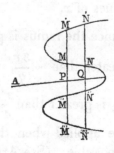

*in a curve, and be cut by the right line* AQ, *in such a manner, that in each case, the sum of the ordinates on one side of* AQ *is equal to the sum of the ordinates on the other, all lines drawn in the curve, parallel to these, will be cut by* AQ *in the same manner.*

Let $y^n - \overline{ax+b}.y^{n-1} + \overline{cx^2+dx+e}.y^{n-2}$ - &c. = 0 be the equation to the curve, reckoning the abscissæ from $A$; also, let $AP = q$, $AQ = r$; then the equation, in the two cases, becomes $y^n - \overline{aq+b}.y^{n-1} + $ &c. = 0, and $y^n - \overline{ar+b}.y^{n-1} + $ &c. = 0; and since in each case, the sum of the positive ordinates is equal to the sum of the negative, $aq + b = 0$, and $ar + b = 0$; and by subtraction $a \times \overline{r-q} = 0$, or $a = 0$; hence, $b = 0$; therefore whatever be the value of $x$, $ax + b = 0$; or the sum of the ordinates on one side of $AQ$ is equal to the sum of the ordinates on the other.

(522.) The line $AQ$ is called a *diameter* of the curve.

(523.) *If the abscissa* APE, *and ordinate* NPQ *cut a curve in as many points as it has dimensions,*

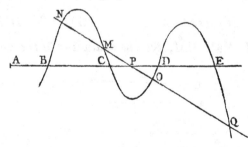

*the rectangle under the segments of the abscissa,* PB × PC × PD × PE, *will be to the rectangle under the ordinates,* PM × PN × PO × PQ, *in an invariable ratio.*

Let

Let $y^n - \overline{ax+b}.y^{n-1} + \&c.\ldots + gx^n + hx^{n-1} + lx^{n-2}$ $+\&c. = 0$ be the equation to the curve; then $gx^n + hx^{n-1} + lx^{n-2} + \&c. = PM \times PN \times PO \times PQ$ (Art. 520); also, the values of $x$, when $y = 0$, are $AB$, $AC$, $AD$, $AE$, that is, the roots of the equation $gx^n + hx^{n-1} + lx^{n-2} + \&c. = 0$, or $x^n + \dfrac{hx^{n-1}}{g} + \dfrac{lx^{n-2}}{g} + \&c. = 0$, are $AB$, $AC$, $AD$, $AE$; and consequently the quantity $x^n + \dfrac{hx^{n-1}}{g} + \dfrac{lx^{n-2}}{g} + \&c. = \overline{AP - AB} \times$

$\overline{AP - AC} \times \overline{AP - AD} \times \overline{AP - AE} = PB \times PC \times PD$ $\times PE$ (Art. 269); and $gx^n + hx^{n-1} + lx^{n-2} + \&c. = g \times PB \times PC \times PD \times PE = PM \times PN \times PO \times PQ$; therefore $PB \times PC \times PD \times PE : PM \times PN \times PO \times PQ$ $:: 1 : g$.

(524.) Cor. If $n = 2$, the curve is a conic section (Art. 501); and if the abscissa be a diameter, or the ordinates on each side of it, $PM$, $PO$, equal to each other, the rectangle under the segments of the abscissa is to the square of the ordinate in an invariable ratio.

(525.) *If there be* n *right lines,* $B\overset{\shortmid\shortmid}{M}$, $C\overset{\shortmid}{M}$, $DM$, &c. *and* $PM$, $P\overset{\shortmid}{M}$, $P\overset{\shortmid\shortmid}{M}$, &c. *be ordinates to the abscissa*

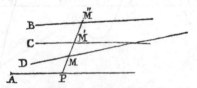

AP, *the relation between the abscissa and ordinates will be expressed by an equation of the form* $y^n - \overline{Ax+B}.y^{n-1} + \overline{Cx^2 + Dx + E}.y^{n-2} - \&c. = 0$.

For

For if $AP=x$, then $PM=ax+b$, $PM'=cx+d$, $PM''=ex+f$, &c. where $a$, $b$, $c$, $d$, $e$, $f$ are invariable (Art. 479); that is, the values of $y$ are $ax+b$, $cx+d$, $ex+f$, &c. therefore $y^n - \overline{a+c+e...} \times \overline{x+b+d+f...} \times y^{n-1} + $ &c. $= 0$, and if $A=a+c+e....$, $B=b+d+f....$, &c. $y^n - \overline{Ax+B}.y^{n-1} + $ &c. $= 0$.

(526.) *If a curve have as many asymptotes as it has dimensions, and a right line be drawn which cuts them all, the parts of the line measured from the asymptotes to the curve, will together be equal to the parts measured, in the same direction, from the curve to the asymptotes.*

Let $y^n - \overline{ax+b}.y^{n-1} + $ &c. $= 0$ be the equation to the curve, and $y^n - \overline{Ax+B}.y^{n-1} + $ &c. $= 0$ the equation to the asymptotes (Art. 525); when $x$ is infinite, the former equation becomes $y^n - axy^{n-1} + $ &c. $= 0$, and the latter $y^n - Axy^{n-1} + $ &c. $= 0$, and these equations coincide (Art. 506), therefore $A=a$; also, $ax+b$ is the sum of the ordinates to the curve, and $Ax+B$, or $ax+B$, is the sum of the ordinates to the asymptotes, in all cases; hence, the difference of these, $b-B$, is an invariable quantity, whatever be the value of $x$; and at an infinite distance this difference is nothing (Art. 506); therefore it is always nothing, or $b=B$; consequently $ax+b=Ax+B$; that is, the sum of the ordinates to the curve is equal to the sum of the ordinates to the asymptotes.

Let $QONM$ be the curve, $AP$ the abscissa, $PQ$ an ordinate, meeting the curve in the points $M$, $N$, $O$, $Q$, and the asymptotes in $a$, $b$, $c$, $d$; then $PM + PN + PO + PQ = Pa + Pb + Pc + Pd$, and by transposition,

position, $PM - Pa + PO - Pc \rightleftharpoons Pb - PN + Pd -$

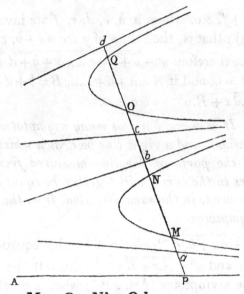

$PQ$, or $aM + cO = Nb + Qd$.

(527.) Cor. In the common hyperbola $MCN$, whose centre is $O$, and asymptotes $Oa$, $Ob$, if any line

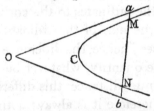

$aMNb$ be drawn cutting the curve in $M$, $N$, and the asymptotes in $a$, $b$, then $aM$ is equal to $Nb$.

(528.) *If a straight line* DC$\overset{\prime}{M}$ *be made to revolve about* C, *and cut the curve* $\overset{\prime}{M}M\overset{\prime\prime}{M}$ *in as many points as it has dimensions; and if* $\dfrac{1}{CD}$ *be always taken equal*

to

*to* $\dfrac{1}{CM}+\dfrac{1}{C\overset{\prime}{M}}-\dfrac{1}{C\overset{\prime\prime}{M}}-\&c.$ *the locus of the point* **D**
*will be a straight line.*

Let *ABP* be an abscissa, and from *M* and *D* draw

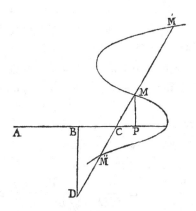

$\overline{MP}$ and $\overline{DB}$ at right angles to **ABP,** and let
$y^{n}-\overline{ax+b}.y^{n-\text{\tiny 1}}\dots\dots+\overline{px^{n-\text{\tiny 1}}+qx^{n-2}+\&c}.y+Px^{n}+$
$Qx^{n-1}+\&c.=0$ be the equation which represents the
relation of *AP* to *PM*. Also, let $AC=z$, $CM=v$,
$CB=w$, $BD=u$; sin. $\angle MCP=s$, sin. $\angle CMP=c$, to
radius 1. Then $1:s::v:sv=y$; and $1:c::v:cv=$
$CP$; hence, $x=AP=z+cv$. If these values of $x$ and
$y$ be substituted in the equation to the curve, the relation of *AC* to *CM* will be known; and the coefficient
of the last term but one of the transformed equation,
divided by the last term, will be the sum of the
reciprocals of it's roots, or $\dfrac{1}{CM}+\dfrac{1}{C\overset{\prime}{M}}-\dfrac{1}{C\overset{\prime\prime}{M}}-\&c.$
(Art. 273). Now, since $x=z+cv$,

$x^{n}\quad=z^{n}\quad+nz^{n-1}cv\quad+n.\dfrac{n-1}{2}z^{n-2}c^{2}v^{2}+\&c.$

$x^{n-1}$

$$x^{n-1} = z^{n-1} + \overline{n-1}.z^{n-2}cv + \overline{n-1}.\frac{n-2}{2}z^{n-3}c^2v^2 + \&c.$$

$$x^{n-2} = z^{n-2} + \overline{n-2}.z^{n-3}cv + \overline{n-.2}.\frac{n-3}{2}z^{n-4}c^2v^2 + \&c.$$

&c.        &c.

and substituting these values for $x$ and it's powers, and $sv$ for $y$, in the terms of the original equation, we have the two last terms of the transformed equation, $\overline{pz^{n-1} + qz^{n-2} + \&c.} \times sv + \overline{nPz^{n-1} + \overline{n-1}.Qz^{n-2} + \&c.}$ $\times cv$, and $Pz^n + Qz^{n-1} + \&c.$ divided, respectively, by $s^n$, all the other terms involving the square or some higher power of $v$; hence,

$$\frac{\overline{s \times pz^{n-1} + qz^{n-2} + \&c.} + c \times \overline{nPz^{n-1} + \overline{n-1}.Qz^{n-2} + \&c.}}{Pz^n + Qz^{n-1} + \&c.} =$$

$$\frac{1}{CM} + \frac{1}{CM'} - \frac{1}{CM''} - \&c. = \frac{1}{CD} = \frac{1}{\sqrt{w^2 + u^2}}, \quad \text{and} \quad s \times$$

$$\overline{pz^{n-1} + qz^{n-2} + \&c.} + c \times \overline{nPz^{n-1} + \overline{n-1}.Qz^{n-2} + \&c.} =$$

$$\frac{1}{\sqrt{w^2 + u^2}} \times \overline{Pz^n + Qz^{n-1} + \&c.}$$

Also, from the similar triangles $MCP$, $BCD$,

$$\sqrt{w^2 + u^2} : u :: 1 : s = \frac{u}{\sqrt{w^2 + u^2}};$$

$$\sqrt{w^2 + u^2} : w :: 1 : c = \frac{w}{\sqrt{w^2 + u^2}};$$

therefore $\dfrac{u}{\sqrt{w^2 + u^2}} \times \overline{pz^{n-1} + qz^{n-2} + \&c.} + \dfrac{w}{\sqrt{w^2 + u^2}}$

$\times \overline{nPz^{n-1} + \overline{n-1}.Qz^{n-2} + \&c.} = \dfrac{1}{\sqrt{w^2 + u^2}} \times$

$\overline{Pz^n + Qz^{n-1} + \&c.}$ or $u \times \overline{pz^{n-1} + qz^{n-2} + \&c.} + w \times$

$\overline{nPz^{n-1} + \overline{n-1}.Qz^{n-2} + \&c.} = Pz^n + Qz^{n-1} + \&c.$ and

since

since the point $C$ is fixed, $AC$, or $z$, is invariable; therefore the relation between $w$ and $u$, or $CB$ and $BD$, is expressed by a simple equation, that is, the locus of the point $D$ is a straight line (Art. 480).

(529.) In the general equation $y^n - \overline{a + bx}.y^{n-1} + \&c. = 0$, if $x$ be so assumed that two roots are impossible, two values of the ordinate belonging to this abscissa are impossible, that is, there are no lines which represent them. Hence it is evident, that in deducing the properties of the ordinates from the equation to the curve, we must suppose all the roots of this equation possible; because, though the sums, powers, products, &c. of such impossible quantities may become possible, and their relations, discovered by an algebraical process, may be expressed by possible quantities, yet the reasoning does not extend to curves, in which the original quantities cannot be represented.

On the subject of Algebraical Curves, the Reader may consult Dr. Waring's *Proprietates Algebraicarum Curvarum,* and Euler's *Anal. Infinitorum.*

THE END.

U

Printed in the United States
By Bookmasters